The Ultimate ECAA Collection

ISBN 978-1-912557-38-7

Published by *RAR Medical Services Limited*
www.uniadmissions.co.uk
info@uniadmissions.co.uk
0208 068 0438

About the Authors

David is a **Merger & Acquisitions Associate** at The Hut Group, a leading online retailer and brand owner in the Beauty & Wellness sectors. Prior to joining The Hut Group, he worked in roles at the Professional Service firm Deloitte, the Investment Bank Greenhill and the Private Equity firm Hgcapital.

David graduated with a **first class honours** in Economics from Gonville and Caius College Cambridge, where he received two college scholarships for outstanding academic performance, in addition to an Essay Prize. He is also a qualified accountant and chartered tax adviser, passing all exams first-time with multiple regional top scores. Since graduating, David has tutored & successfully provided academic coaching to hundreds of students, both in a personal capacity and for university admissions.

Rohan is the **Director of Operations** at *UniAdmissions* and is responsible for its technical and commercial arms. He graduated from Gonville and Caius College, Cambridge and is a fully qualified doctor. Over the last five years, he has tutored hundreds of successful Oxbridge and Medical applicants. He has also authored ten books on admissions tests and interviews.

Rohan has taught physiology to undergraduates and interviewed medical school applicants for Cambridge. He has published research on bone physiology and writes education articles for the Independent and Huffington Post. In his spare time, Rohan enjoys playing the piano and table tennis.

The Ultimate ECAA Collection

Three Books in One

David Meacham
Rohan Agarwal

How to use this Book

Congratulations on taking the first step to your ECAA preparation! First used in 2008, the ECAA is a difficult exam and you'll need to prepare thoroughly in order to make sure you get that dream university place.

The *Ultimate ECAA Collection* is the most comprehensive ECAA book available – it's the culmination of three top-selling ECAA books:

- *The Ultimate ECAA Guide*
- *ECAA Past Paper Solutions*
- *ECAA Practice Papers*

Whilst it might be tempting to dive straight in with mock papers, this is not a sound strategy. Instead, you should approach the ECAA in the three steps shown below. Firstly, start off by understanding the structure, syllabus and theory behind the test. Once you're satisfied with this, move onto doing the 300 practice questions found in *The Ultimate ECAA Guide* (not timed!). Then, once you feel ready for a challenge, do each past paper under timed conditions. Start with the 2016 paper and work chronologically; check your solutions against the model answers given in *ECAA Past Paper Worked Solutions*. Finally, once you've exhausted these, go through the two ECAA Mock Papers found in *ECAA Practice Papers* – these are a final boost to your preparation.

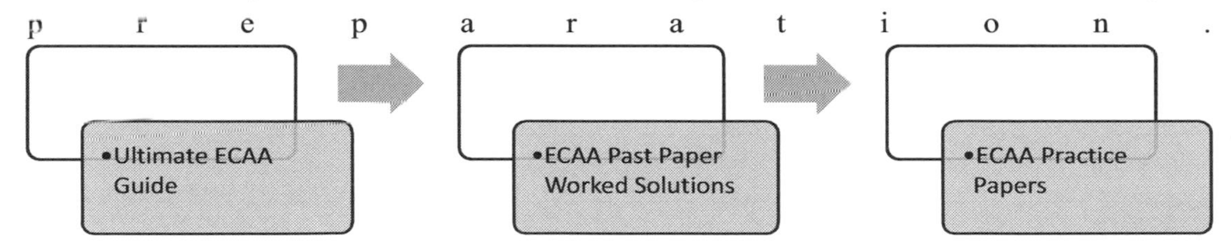

As you've probably realised by now, there are well over 500 questions to tackle meaning that this isn't a test that you can prepare for in a single week. From our experience, the best students will prepare anywhere between four to eight weeks (although there are some notable exceptions!).

Remember that the route to a high score is your approach and practice. Don't fall into the trap that "you can't prepare for the ECAA"– this could not be further from the truth. With knowledge of the test, some useful time-saving techniques and plenty of practice you can dramatically boost your score.

Work hard, never give up and do yourself justice. Good luck!

The Ultimate ECAA Guide

What is the ECAA?
The Economics Admissions Assessment (ECAA) is a two hour written exam for prospective Cambridge Economics applicants.

What does the ECAA consist of?

Section	SKILLS TESTED	Questions	Timing
1A 1B	Problem-solving Advanced Mathematics	20 MCQs 15 MCQs	80 minutes
2	Writing Task	One Long Essay	40 minutes

Why is the ECAA used?
Cambridge applicants tend to be a bright bunch and therefore usually have excellent grades. The majority of economics applicants score in excess of 90% in their A level subjects. This means that competition is fierce – meaning that the universities must use the ECAA to help differentiate between applicants.

When do I sit ECAA?
The ECAA normally takes place in the first week of November every year, normally on a Wednesday Morning.

Can I resit the ECAA?
No, you can only sit the ECAA once per admissions cycle.

Where do I sit the ECAA?
You can usually sit the ECAA at your school or college (ask your exams officer for more information). Alternatively, if your school isn't a registered test centre or you're not attending a school or college, you can sit the ECAA at an authorised test centre.

Who has to sit the ECAA?
All applicants for Cambridge Economics need to sit the test.

Do I have to resit the ECAA if I reapply?
Yes, each admissions cycle is independent - you cannot use your score from any previous attempts.

How is the ECAA Scored?
In section 1, each question carries one mark and there is no negative marking. Both sections 1A + 1B are equally weighted. In section 2, your answer will be assessed based on the argument and also its clarity.

How is the ECAA used?
Different Cambridge colleges will place different weightings on different components so its important you find out as much information about how your marks will be used by emailing the college admissions office.

In general, the university will interview a high proportion of realitstic applicants so the ECAA score isn't vital for making the interview shortlist. However, it can play a huge role in the final decision after your interview

General Advice

Start Early

It is much easier to prepare if you practice little and often. Start your preparation well in advance; ideally by mid September but at the latest by early October. This way you will have plenty of time to complete as many papers as you wish to feel comfortable and won't have to panic and cram just before the test, which is a much less effective and more stressful way to learn. In general, an early start will give you the opportunity to identify the complex issues and work at your own pace.

Prioritise

Some questions in sections 1 + 2 can be long and complex – and given the intense time pressure you need to know your limits. It is essential that you don't get stuck with very difficult questions. If a question looks particularly long or complex, mark it for review and move on. You don't want to be caught 5 questions short at the end just because you took more than 3 minutes in answering a challenging multi-step physics question. If a question is taking too long, choose a sensible answer and move on. Remember that each question carries equal weighting and therefore, you should adjust your timing in accordingly. With practice and discipline, you can get very good at this and learn to maximise your efficiency.

Positive Marking

There are no penalties for incorrect answers in the ECAA; you will gain one for each right answer and will not get one for each wrong or unanswered one. This provides you with the luxury that you can always guess should you absolutely be not able to figure out the right answer for a question or run behind time. Since each question provides you with 4 to 6 possible answers, you have a 16-25% chance of guessing correctly. Therefore, if you aren't sure (and are running short of time), then make an educated guess and move on. Before 'guessing' you should try to eliminate a couple of answers to increase your chances of getting the question correct. For example, if a question has 5 options and you manage to eliminate 2 options- your chances of getting the question increase from 20% to 33%!

Avoid losing easy marks on other questions because of poor exam technique. Similarly, if you have failed to finish the exam, take the last 10 seconds to guess the remaining questions to at least give yourself a chance of getting them right.

Practice

This is the best way of familiarising yourself with the style of questions and the timing for this section. Practising questions will put you at ease and make you more comfortable with the exam. The more comfortable you are, the less you will panic on the test day and the more likely you are to score highly. Initially, work through the questions at your own pace, and spend time carefully reading the questions and looking at any additional data. When it becomes closer to the test, **make sure you practice the questions under exam conditions**.

Past Papers

The ECAA is a very new exam so there aren't many sample papers available. Specimen papers are freely available online at www.uniadmissions.co.uk/ECAA. Once you've worked your way through the questions in this book, you are highly advised to attempt them.

Repeat Questions

When checking through answers, pay particular attention to questions you have got wrong. If there is a worked answer, look through that carefully until you feel confident that you understand the reasoning, and then repeat the question without help to check that you can do it. If only the answer is given, have another look at the question and try to work out why that answer is correct. This is the best way to learn from your mistakes, and means you are less likely to make similar mistakes when it comes to the test. The same applies for questions which you were unsure of and made an educated guess which was correct, even if you got it right. When working through this book, **make sure you highlight any questions you are unsure of**, this means you know to spend more time looking over them once marked.

Top tip! In general, students tend to improve the fastest in section 2 and slowest in section 1A; section

No Calculators

You aren't permitted to use calculators in the ECAA – thus, it is essential that you have strong numerical skills. For instance, you should be able to rapidly convert between percentages, decimals and fractions. You will seldom get questions that would require calculators but you would be expected to be able to arrive at a sensible estimate. Consider for example:

Estimate 3.962 x 2.322;

3.962 is approximately 4 and 2.323 is approximately 2.33 = 7/3.

Thus, 3.962 x 2.322 $4\,x\frac{7}{3} = \frac{28}{3} = 9.33$

Since you will rarely be asked to perform difficult calculations, you can use this as a signpost of if you are tackling a question correctly. For example, when solving a physics question, you end up having to divide 8,079 by 357- this should raise alarm bells as calculations in the ECAA are rarely this difficult.

A word on timing...

"If you had all day to do your ECAA, you would get 100%. But you don't."

Whilst this isn't completely true, it illustrates a very important point. Once you've practiced and know how to answer the questions, the clock is your biggest enemy. This seemingly obvious statement has one very important consequence. **The way to improve your ECAA score is to improve your speed.** There is no magic bullet. But there are a great number of techniques that, with practice, will give you significant time gains, allowing you to answer more questions and score more marks.

Timing is tight throughout the ECAA – **mastering timing is the first key to success**. Some candidates choose to work as quickly as possible to save up time at the end to check back, but this is generally not the best way to do it. ECAA questions can have a lot of information in them – each time you start answering a question it takes time to get familiar with the instructions and information. By splitting the question into two sessions (the first run-through and the return-to-check) you double the amount of time you spend on familiarising yourself with the data, as you have to do it twice instead of only once. This costs valuable time. In addition, candidates who do check back may spend 2–3 minutes doing so and yet not make any actual changes. Whilst this can be reassuring, it is a false reassurance as it is unlikely to have a significant effect on your actual score. Therefore it is usually best to pace yourself very steadily, aiming to spend the same amount of time on each question and finish the final question in a section just as time runs out. This reduces the time spent on re-familiarising with questions and maximises the time spent on the first attempt, gaining more marks.

It is essential that you don't get stuck with the hardest questions – no doubt there will be some. In the time spent answering only one of these you may miss out on answering three easier questions. If a question is taking too long, choose a sensible answer and move on. Never see this as giving up or in any way failing, rather it is the smart way to approach a test with a tight time limit. With practice and discipline, you can get very good at this and learn to maximise your efficiency. It is not about being a hero and aiming for full marks – this is almost impossible and very much unnecessary (even Oxbridge will regard any score higher than 7 as exceptional). It is about maximising your efficiency and gaining the maximum possible number of marks within the time you have.

Top tip! Ensure that you take a watch that can show you the time in seconds into the exam. This will allow you have a much more accurate idea of the time you're spending on a question. In general, if

Use the Options:

Some questions may try to overload you with information. When presented with large tables and data, it's essential you look at the answer options so you can focus your mind. This can allow you to reach the correct answer a lot more quickly. Consider the example below:

The table below shows the results of a study investigating antibiotic resistance in staphylococcus populations. A single staphylococcus bacterium is chosen at random from a similar population. Resistance to any one antibiotic is independent of resistance to others.

Antibiotic	Number of Bacteria tested	Number of Resistant Bacteria
Benzyl-penicillin	10^{11}	98
Chloramphenicol	10^{9}	1200
Metronidazole	10^{8}	256
Erythromycin	10^{5}	2

Calculate the probability that the bacterium selected will be resistant to all four drugs.

A. 1 in 10^{6}
B. 1 in 10^{12}
C. 1 in 10^{20}
D. 1 in 10^{25}
E. 1 in 10^{30}
F. 1 in 10^{35}

Looking at the options first makes it obvious that there is **no need to calculate exact values**- only in powers of 10. This makes your life a lot easier. If you hadn't noticed this, you might have spent well over 90 seconds trying to calculate the exact value when it wasn't even being asked for.

In other cases, you may actually be able to use the options to arrive at the solution quicker than if you had tried to solve the question as you normally would. Consider the example below:

A region is defined by the two inequalities: $x - y^2 > 1$ and $xy > 1$. Which of the following points is in the defined region?

A. (10,3)
B. (10,2)
C. (-10,3)
D. (-10,2)
E. (-10,-3)

Whilst it's possible to solve this question both algebraically or graphically by manipulating the identities, by far **the quickest way is to actually use the options.** Note that options C, D and E violate the second inequality, narrowing down to answer to either A or B. For A: $10 - 3^2 = 1$ and thus this point is on the boundary of the defined region and not actually in the region. Thus the answer is B (as $10-4 = 6 > 1$.)

In general, it pays dividends to look at the options briefly and see if they can be help you arrive at the question more quickly. Get into this habit early – it may feel unnatural at first but it's guaranteed to save you time in the long run.

Keywords

If you're stuck on a question; pay particular attention to the options that contain key modifiers like "**always**", "**only**", "**all**" as examiners like using them to test if there are any gaps in your knowledge. E.g. the statement "arteries carry oxygenated blood" would normally be true; "All arteries carry oxygenated blood" would be false because the pulmonary artery carries deoxygenated blood.

SECTION 1

This is the first section of the ECAA and as you walk in, it is inevitable that you will feel nervous. Make sure that you have been to the toilet because once it starts you cannot simply pause and go. Take a few deep breaths and calm yourself down. Remember that panicking will not help and may negatively affect your marks- so try and avoid this as much as possible.

You have one hour to answer 35 questions in section 1. Whilst this section of the ECAA is renowned for being difficult to prepare for, there are powerful shortcuts and techniques that you can use to save valuable time on these types of questions.

You have just above 2 minutes per question; this may sound like a lot but given that you're often required to analyse passages or graphs- it can often not be enough. Some questions in this section are very tricky and can be a big drain on your limited time. **The people who fail to complete section 1 are those who get bogged down on a particular question**.

Therefore, it is vital that you start to get a feel for which questions are going to be easy and quick to do and which ones should be left till the end. The best way to do this is through practice and the questions in this book will offer extensive opportunities for you to do so.

SECTION 1A: Problem Solving Questions

Section 1 problem solving questions are arguably the hardest to prepare for. However, there are some useful techniques you can employ to solve some types of questions much more quickly:

Construct Equations

Some of the problems in Section 1 are quite complex and you'll need to be comfortable with turning prose into equations and manipulating them. For example, when you read "Mark is twice as old as Jon" – this should immediately register as M = 2J. Once you get comfortable forming equations, you can start to approach some of the harder questions in this book (and past papers) which may require you to form and solve simultaneous equations. Consider the example:

Nick has a sleigh that contains toy horses and clowns and counts 44 heads and 132 legs in his sleigh. Given that horses have one head and four legs, and clowns have one head and two legs, calculate the difference between the number of horses and clowns.

A. 0

B. 5

C. 22

D. 28

E. 132

F. More information is needed.

To start with, let C= Clowns and H= Horses.
For Heads: $C + H = 44$; For Legs: $2C + 4H = 132$
This now sets up your two equations that you can solve simultaneously.
$C = 44 - H$ so $2(44 - H) + 4H = 132$
Thus, $88 - 2H + 4H = 132$;
Therefore, $2H = 44$; $H = 22$
Substitute back in to give $C = 44 - H = 44 - 22 = 22$
Thus the difference between horses and clowns $= C - H = 22 - 22 = 0$

It's important you are able to do these types of questions quickly (and **without resorting to trial & error** as they are commonplace in section 1.

Spatial Reasoning

There are usually 1-2 spatial reasoning questions every year. They usually give nets for a shape or a patterned cuboid and ask which options are possible rotations. Unfortunately, they are extremely difficult to prepare for because the skills necessary to solve these types of questions can take a very long time to improve. The best thing you can do to prepare is to familiarise yourself with the basics of how cube nets work and what the effect of transformations are e.g. what happens if a shape is reflected in a mirror etc.

It is also a good idea to try to learn to draw basic shapes like cubes from multiple angles if you can't do so already. Finally, remember that if the shape is straightforward like a cube, it might be easier for you to draw a net, cut it out and fold it yourself to see which of the options are possible.

Diagrams

When a question asks about timetables, orders or sequences, draw out diagrams. By doing this, you can organise your thoughts and help make sense of the question.

"Mordor is West of Gondor but East of Rivendale. Lorien is midway between Gondor and Mordor. Erebus is West of Mordor. Eden is not East of Gondor."

*Which of the following **cannot** be concluded?*

A. Lorien is East of Erebus and Mordor.

B. Mordor is West of Gondor and East of Erebus.

C. Rivendale is west of Lorien and Gondor.

D. Gondor is East of Mordor and East of Lorien

E. Erebus is West of Mordor and West of Rivendale.

Whilst it is possible to solve this in your head, it becomes much more manageable if you draw a quick diagram and plot the positions of each town:

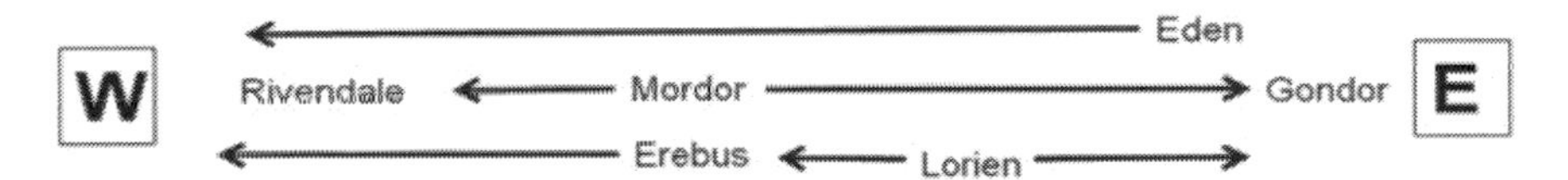

Now, it's a simple case of going through each option and seeing if it is correct according to the diagram. You can now easily see that Option E- Erebus cannot be west of Rivendale.

Don't feel that you have to restrict yourself to linear diagrams like this either – for some questions you may need to draw tables or even Venn diagrams. Consider the example:

Slifers and Osiris are not legendary. Krakens and Minotaurs are legendary. Minotaurs and Lords are both divine. Humans are neither legendary nor divine.

A. Krakens may be only legendary or legendary and divine.
B. Humans are not divine.
C. Slifers are only divine.
D. Osiris may be divine.
E. Humans and Slifers are the same in terms of both qualities.

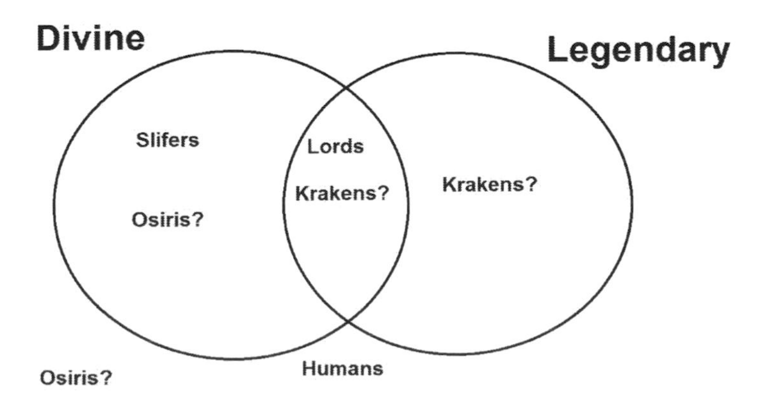

Constructing a Venn diagram allows us to quickly see that the position of Osiris and Krakens aren't certain. Thus, A and D must be true. Humans are neither so B is true. Krakens may be divine so A is true. E cannot be concluded as Slifers are divine but are humans are not. Thus, E is False.

Problem Solving Questions

Question 1:
The hospital coordinator is making the rota for the ward for next week; two of Drs Evans, James and Luca must be working on weekdays, none of them on Sundays and all of them on Saturdays. Dr Evans works 4 days a week including Mondays and Fridays. Dr Luca cannot work Monday or Thursday. Only Dr James can work 4 days consecutively, but he cannot do 5.

What days does Dr James work?
A. Saturday, Sunday and Monday.
B. Monday, Tuesday, Wednesday, Thursday and Saturday.
C. Monday, Thursday Friday and Saturday.
D. Tuesday, Wednesday, Friday and Saturday.
E. Monday, Tuesday, Wednesday, Thursday and Friday.

Question 2:
Michael, a taxi driver, charges a call out rate and a rate per mile for taxi rides. For a 4 mile ride he charges £11, and for a 5 mile ride, £13.

How much does he charge for a 9-mile ride?

A. £15
B. £17
C. £19
D. £20
E. £21

Question 3:
Goblins and trolls are not magical. Fairies and goblins are both mythical. Elves and fairies are magical. Gnomes are neither mythical nor magical.

Which of the following is **FALSE**?
A. Elves may be only magical or magical and mythical.
B. Gnomes are not mythical.
C. Goblins are only mythical.
D. Trolls may be mythical.
E. Gnomes and goblins are the same in terms of both qualities.

Question 4:
Jessica runs a small business making bespoke wall tiles. She has just had a rush order for 100 tiles placed that must be ready for today at 7pm. The client wants the tiles packed all together, a process which will take 15 minutes. Only 50 tiles can go in the kiln at any point and they must be put in the kiln to heat for 45 minutes. The tiles then sit in the kiln to cool before they can be packed, a process which takes 20 minutes. While tiles are in the kiln Jessica is able to decorate more tiles at a rate of 1 tile per minute.

What is the latest time Jessica can start making the tiles?

A. 2:55pm
B. 3:15pm
C. 3:30pm
D. 3:45pm

Question 5:
Pain nerve impulses are twice as fast as normal touch impulses. If Yun touches a boiling hot pan this message reaches her brain, 1 metre away, in 1 millisecond.

What is the speed of a normal touch impulse?

A. 5 m/s
B. 20 m/s
C. 50 m/s
D. 200m/s
E. 500 m/s

Question 6:

A woman has two children Melissa and Jack, yearly, their birthdays are 3 months apart, both being on the 22nd. The woman wishes to continue the trend of her children's names beginning with the same letter as the month they were born. If her next child, Alina is born on the 22nd 2 months after Jack's birthday, how many months after Alina is born will Melissa have her next birthday?

A. 2 months
B. 4 months
C. 5 months
D. 6 months
E. 7 months

Question 7:

Policemen work in pairs. PC Carter, PC Dirk, PC Adams and PC Bryan must work together but not for more than seven days in a row, which PC Adams and PC Bryan now have. PC Dirk has worked with PC Carter for 3 days in a row. PC Carter does not want to work with PC Adams if it can be avoided.

Who should work with PC Bryan?

A. PC Carter
B. PC Dirk
C. PC Adams
D. Nobody is available under the guidelines above.

Question 8:

My hair-dressers charges £30 for a haircut, £50 for a cut and blow-dry, and £60 for a full hair dye. They also do manicures, of which the first costs £15, and includes a bottle of nail polish, but are subsequently reduced by £5 if I bring my bottle of polish. The price is reduced by 10% if I book and pay for the next 5 appointments in advance and by 15% if I book at least the next 10.

I want to pay for my next 5 cut and blow-dry appointments, as well as for my next 3 manicures. How much will it cost?

A. £170
B. £255
C. £260
D. £285
E. £305

Question 9:

Alex, Bertha, David, Gemma, Charlie, Elena and Frankie are all members of the same family consisting of three children, two of whom, Frankie and Gemma are girls. No other assumption of gender based on name can be established. There are also four adults. Alex is a doctor and is David's brother. One of them is married to Elena, and they have two children. Bertha is married to David; Gemma is their child.

Who is Charlie?

A. Alex's daughter
B. Frankie's father
C. Gemma's brother
D. Elena's son
E. Gemma's sister

Question 10:

At 14:30 three medical students were asked to examine a patient's heart. Having already watched their colleague, the second two students were twice as fast as the first to examine. During the 8 minutes break after the final student had finished, they were told by their consultant that they had taken too long and so should go back and do the examinations again. The second time all the students took half as long as they had taken the first time with the exception of the first student who, instead took the same time as his two colleagues' second attempt. Assuming there was a one minute change over time between each student and they were finished by 15:15, how long did the second student take to examine the first time?

A. 3 minutes
B. 4 minutes
C. 6 minutes
D. 7 minutes
E. 8 minutes

Question 11:

I pay for 2 chocolate bars that cost £1.65 each with a £5 note. I receive 8 coins change, only 3 of which are the same.

Which **TWO** coins do I not receive in my change?

A. 1p
B. 2p
C. 5p
D. 10p
E. 20p
F. £2
G. £1

Question 12:

Two 140m long trains are running at the same speed in opposite directions. If they cross each other in 14 seconds then what is speed of each train?

A. 10 km/hr
B. 18 km/hr
C. 32 km/hr
D. 36 km/hr
E. 42 km/hr

Question 13:

Anil has to refill his home's swimming pool. He has four hoses which all run at different speeds. Alone, the first would completely fill the pool with water in 6 hours, the second in two days, the third in three days and the fourth in four days.

Using all the hoses together, how long will it take to fill the pool to the nearest quarter of an hour?

A. 4 hours 15 minutes
B. 4 hours 30 minutes
C. 4 hours 45 minutes
D. 5 hours
E. 5 hours 15 minutes

Question 14:

An ant is stuck in a 30 cm deep ditch. When the ant reaches the top of the ditch he will be able to climb out straight away. The ant is able to climb 3 cm upwards during the day, but falls back 2 cm at night.

How many days does it take for the ant to climb out of the ditch?

A. 27
B. 28
C. 29
D. 30
E. 31

Question 15:

When buying his ingredients a chef gets a discount of 10% when he buys 10 or more of each item, and 20% discount when he buys 20 or more. On one order he bought 5 sausages and 10 Oranges and paid £8.50. On another, he bought 10 sausages and 10 apples and paid £9, on a third he bought 30 oranges and paid £12.

How much would an order of 2 oranges, 13 sausages and 12 apples cost?

A. £12.52
B. £12.76
C. £13.52
D. £13.76
E. £13.80

Question 16:

My hairdressers encourage all of its clients to become members. By paying an annual member fee, the cost of haircuts decreases. VIP membership costs £125 annually with a £10 reduction on haircuts. Executive VIP membership costs £200 for the year with a £15 reduction per haircut. At the moment I am not a member and pay £60 per haircut. I know how many haircuts I have a year, and I work out that by becoming a member on either programme it would work out cheaper, and I would save the same amount of money per year on either programme.

How much will I save this year by buying membership?

A. £10
B. £15
C. £25
D. £30
E. £50

Question 17:

If criminals, thieves and judges are represented below:

Criminals **Thieves** **Judges**

Assuming that judges must have clean record, all thieves are criminals and all those who are guilty are convicted of their crimes, which of one of the following best represents their interaction?

A. 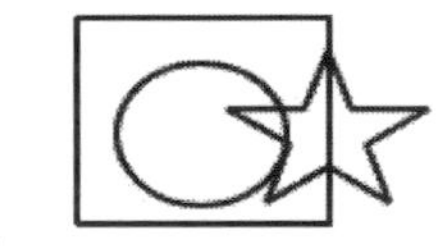

B. 

C. 

D. 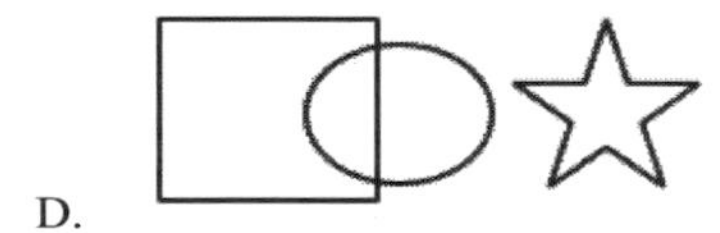

E. 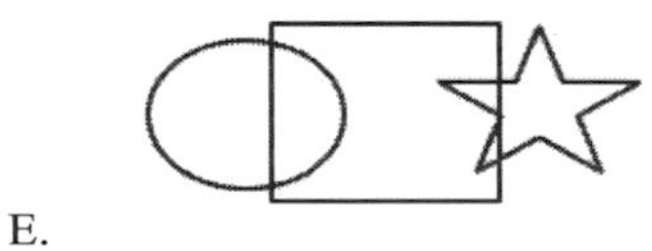

Question 18:

The months of the year have been made into number codes. The code is comprised of three factors, including two of these being related the letters that make up the name of the month. No two months would have the same first number. But some such as March, which has the code 3513, have the same last number as others, such as May, which has the code 5313. October would be coded as 10715 while February is 286.

What would be the code for April?

A. 154
B. 441
C. 451
D. 514
E. 541

Question 19:

A mother gives yearly birthday presents of money to her children based on the age and their exam results. She gives them £5 each plus £3 for every year they are older than 5, and a further £10 for every A* they achieved in their results. Josie is 16 and gained 9 A*s in her results. Although Josie's brother Carson is 2 years older he receives £44 less a year for his birthday.

How many more A*s did Josie get than Carson?

A. 2
B. 3
C. 4
D. 5
E. 10

Question 20:

Apples are more expensive than pears, which are more expensive than oranges. Peaches are more expensive than oranges. Apples are less expensive than grapes.

Which two of the following must be true?

A. Grapes are less expensive than oranges.
B. Peaches may be less expensive than pears.
C. Grapes are more expensive than pears.
D. Pears and peaches are the same price.
E. Apples and peaches are the same price.

Question 21:

What is the minimum number of straight cutting motions needed to slice a cylindrical cake into 8 equally sized pieces?

A. 2
B. 3
C. 4
D. 5
E. 6
F. 8

Question 22:

Three friends, Mark, Russell and Tom had agreed to meet for lunch at 12 PM on Sunday. Daylight saving time (GMT+1) had started at 2 AM the same day, where clocks should be put forward by one hour. Mark's phone automatically changes the time but he does not realise this so when he wakes up he puts his phone forward an hour and uses his phone to time his arrival to lunch. Tom puts all of his clocks forward one hour at 7 AM. Russell forgets that the clocks should go forward, wakes at 10 AM doesn't change his clocks. All of the friends arrive on time as far as they are concerned.

Assuming that none of the friends realise any errors before arriving, which **TWO** of the following statements are **FALSE**?

A. Tom arrives at 12 PM (GMT +1).
B. All three friends arrive at the same time.
C. There is a 2 hour difference between when the first and last friend arrive.
D. Mark arrives late.
E. Mark arrives at 1 PM (GMT+3).
F. Russell arrives at 12 PM (GMT+0).

Question 23:

A class of young students has a pet spider. Deciding to play a practical joke on their teacher, one day during morning break one of the students put the spider in their teachers' desk. When first questioned by the head teacher, Mr Jones, the five students who were in the classroom during morning break all lied about what they saw. Realising that the students were all lying, Mr Jones called all 5 students back individually and, threatened with suspension, all the students told the truth. Unfortunately Mr Jones only wrote down the student's statements not whether they had been told in the truthful or lying questioning.

The students' two statements appear below:

Archie: "It wasn't Edward. "
"It was Bella."

Darcy: "It was Charlotte"
"It was Bella"

Edward: "It was Darcy"
"It wasn't Archie"

Charlotte: "It was Edward."
"It wasn't Archie"

Bella: "It wasn't Charlotte."
"It wasn't Edward."

Who put the spider in the teacher's desk?

A. Edward
B. Bella
C. Darcy
D. Charlotte
E. More information needed.

Question 24:
Dr Massey wants to measure out 0.1 litres of solution. Unfortunately the lab assistant dropped the 200 ml measuring cylinder, and so the scientist only has a 300 ml and a half litre-measuring beaker. Assuming he cannot accurately use the beakers to measure anything less than their full capacity, what is the minimum volume he will have to use to be able to ensure he measures the right amount?

A. 100 ml
B. 200 ml
C. 300 ml
D. 400 ml
E. 500 ml
F. 600 ml

Question 25:
Francis lives on a street with houses all consecutively numbered evenly. When one adds up the value of all the house numbers it totals 870.

In order to determine Francis' house number:
1. The relative position of Francis' house must be known.
2. The number of houses in the street must be known.
3. At least three of the house numbers must be known.

A. 1 only
B. 2 only
C. 3 only
D. 1 and 2
E. 2 and 3

Question 26:
There were 20 people exercising in the cardio room of a gym. Four people were about to leave when suddenly a man collapsed on one of the machines. Fortunately a doctor was on the machine beside him. Emerging from his office, one of the personal trainers called an ambulance. In the 5 minutes that followed before the two paramedics arrived, half of the people who were leaving, left upon hearing the commotion, and eight people came in from the changing rooms to hear the paramedics pronouncing the man dead.

How many living people were left in the room?

A. 25
B. 26
C. 27
D. 28
E. 29
F. 30

Question 27:
A man and woman are in an accident. They both suffer the same trauma, which causes both of them to lose blood at a rate of 0.2 Litres/minute. At normal blood volume the man has 8 litres and the woman 7 litres, and people collapse when they lose 40% of their normal blood volume.

Which **TWO** of the following are true?

A. The man will collapse 2 minutes before the woman.
B. The woman collapses 2 minutes before the man.
C. The total blood loss is 5 litres.
D. The woman has 4.2 litres of blood in her body when she collapses.
E. The man's blood loss is 4.8 litres when he collapses.
F. Blood loss is at a rate of 2 litres every 12 minutes.

Question 28:
Jenny, Helen and Rachel have to run a distance of 13 km. Jenny runs at a pace of 8 kmph, Helen at a pace of 10 kmph, and Rachel 11 kmph.

If Jenny sets off 15 minutes before Helen, and 25 minutes before Rachel, what order will they arrive at the destination?

A. Jenny, Helen, Rachel.
B. Helen, Rachel, Jenny.
C. Helen, Jenny, Rachel.
D. Rachel, Helen, Jenny.
E. Jenny, Rachel, Helen.
F. None of the above.

Question 29:
On a specific day at a GP surgery 150 people visited the surgery and common complaints were recorded as a percentage of total patients. Each patient could use their appointment to discuss up to 2 complaints. 56% flu-like symptoms, 48% pain, 20% diabetes, 40% asthma or COPD, 30% high blood pressure.

Which statement **must** be true?

A. A minimum of 8 patients complained of pain and flu-like symptoms.
B. No more than 45 patients complained of high blood pressure and diabetes.
C. There were a minimum of 21 patients who did not complain about flu-like symptoms or high blood pressure.
D. There were actually 291 patients who visited the surgery.
E. None of the above.

Question 30:
All products in a store were marked up by 15%. They were subsequently reduced in a sale with quoted saving of 25% from the higher price. What is the true reduction from the original price?

A. 5%
B. 10%
C. 13.75%
D. 18.25%
E. 20%
F. None of the above.

Question 31:

A recipe states it makes 12 pancakes and requires the following ingredients: 2 eggs, 100g plain flour, and 300ml milk. Steve is cooking pancakes for 15 people and wants to have sufficient mixture for 3 pancakes each.

What quantities should Steve use to ensure this whilst using whole eggs?

A. 2½ eggs, 125g plain flour, 375ml milk
B. 3 eggs , 150g plain flour, 450 ml milk
C. 7½ eggs, 375g plain flour, 1125 ml milk
D. 8 eggs, 400g plain flour, 1200 ml milk
E. 12 eggs, 600g plain flour, 1800 ml milk
F. None of the above.

Question 32:

Spring Cleaning cleaners buy industrial bleach from a warehouse and dilute it twice before using it domestically. The first dilution is by 9:1 and then the second, 4:1.

If the cleaners require 6 litres of diluted bleach, how much warehouse bleach do they require?

A. 30 ml
B. 120 ml
C. 166 ml
D. 666 ml
E. 1,200 ml
F. None of the above

Question 33:

During a GP consultation in 2015, Ms Smith tells the GP about her grandchildren. Ms Smith states that Charles is the middle grandchild and was born in 2002. In 2010, Bertie was twice the age of Adam and that in 2015 there are 5 years between Bertie and Adam. Charles and Adam are separated by 3 years.

How old are the 3 grandchildren in 2015?

A. Adam = 16, Bertie = 11, Charles = 13
B. Adam = 5, Bertie = 10, Charles = 8
C. Adam = 10, Bertie = 15, Charles = 13
D. Adam = 10, Bertie = 20, Charles = 13
E. Adam = 11, Bertie = 10, Charles = 8
F. More information needed.

Question 34:

Kayak Hire charges a fixed flat rate and then an additional half-hourly rate. Peter hires the kayak for 3 hours and pays £14.50, and his friend Kevin hires 2 kayaks for 4hrs30mins each and pays £41. How much would

Tom pay to hire one kayak for 2 hours?

A. £8
B. £10.50
C. £15
D. £33.20
E. £35.70
F. None of the above.

Question 35:

A ticketing system uses a common digital display of numbers 0 – 9. The number 7 is showing. However, a number of the light elements are not currently working.

Which set of the following digits is possible?

A. 3, 4, 7
B. 0, 1, 9
C. 2, 7, 8
D. 0, 5, 9
E. 3, 8, 9
F. 3, 4, 9

Question 36:

A team of 4 builders take 12 days of 7 hours work to complete a house. The company decides to recruit 3 extra builders.

How many 8 hour days will it take the new workforce to build a house?

A. 2 days
B. 6 days
C. 7 days
D. 10 days
E. 12 days
F. More information needed

Question 37:

All astragalus are fabacaea as are all gummifer. Acacia are not astragalus. Which of the following statements is true?

A. Acacia are not fabacaea.
B. No astragalus are also gummifer.
C. All fabacae are astragalus or gummifer.
D. Some acacia may be fabacaea.
E. Gummifer are all acacia.
F. None of the above.

Question 38:

The Smiths want to reupholster both sides of their seating cushions (dimensions shown on diagram). The fabric they are using costs £10/m, can only be bought in whole metre lengths and has a standard width of 1m. Each side of a cushion must be made from a single piece of fabric. The seamstress changes a flat rate of £25 per cushion. How much will it cost them to reupholster 4 cushions?

A. £ 20
B. £ 80
C. £ 110
D. £ 130
E. £ 150
F. £ 200

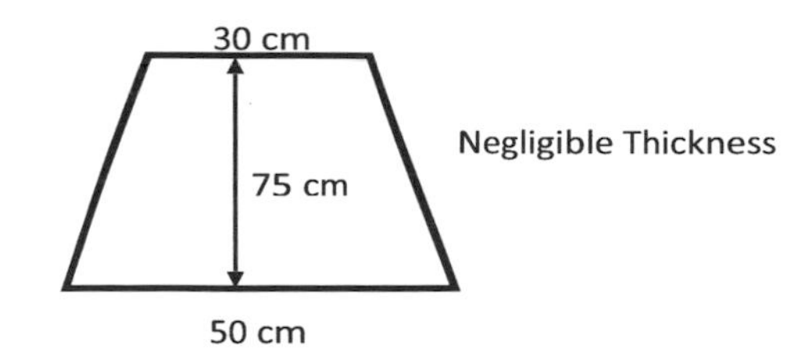

Question 39:

Lisa buys a cappuccino from either Milk or Beans Coffee shops each day. The quality of the coffee is the same but she wishes to work out the relative costs once the loyalty scheme has been taken into account. In Milk, a regular cappuccino is £2.40, and in Beans, £2.15. However, the loyalty scheme in Milk gives Lisa a free cappuccino for every 9 she buys, whereas Beans use a points system of 10 points per full pound spent (each point is worth 1p) which can be used to cover the cost of a full cappuccino.

If Lisa buys a cappuccino each day of September, which coffee shop would work out cheaper, and by how much?

A. Milk, by £4.60
B. Beans by £6.30
C. Beans, by £4.60
D. Beans, by £2.45
E. Milk, by £2.45
F. Milk, by £6.25

Question 40:

Paula needs to be at a meeting in Notting Hill at 11am. The route requires her to walk 5 minutes to the 283 bus which takes 25 minutes, and then change to the 220 bus which takes 14 minutes. Finally she walks for 3 minutes to her meeting. If the 283 bus comes every 10 minutes, and the 220 bus at 0 minutes, 20 minutes and 40 minutes past the hour, what is the latest time she can leave and still be at her meeting on time?

A. 09.45
B. 09.58
C. 10.01
D. 10.05
E. 10.10
F. 10.15

Question 41:

Two trains, a high speed train A and a slower local train B, travel from Manchester to London. Train A travels the first 20 km at 100 km/hr and then at an average speed of 150km/hr. Train B travels at a constant average speed of 90 km/hr. If train B leaves 20 minutes before train A, at what distance will train A pass train B?

A. 75 km
B. 90 km
C. 100 km
D. 120 km
E. 150 km

Question 42:

The university gym has an upfront cost of £35 with no contract fee, but classes are charged at £3 each. The local gym has no joining fee and is £15 per month. What is the minimum number of classes I need to attend in a 12 month period to make the local gym cheaper than the university gym?

A. 40
B. 48
C. 49
D. 50
E. 55
F. 60

Question 43:

"All medicines are drugs, but not all drugs are medicines", goes a well-known saying. If we accept this statement as true, and consider that all antibiotics are medicines, but no herbal drugs are medicines, then which of the following is definitely **FALSE**?

A. Some herbal drugs are not medicines.
B. All antibiotics are drugs.
C. Some herbal drugs are antibiotics.
D. Some medicines are antibiotics

Question 44:

Sonia has been studying the paths taken by various trains travelling between London and Edinburgh on the East coast. Trains can stop at the following stations: Newark, Peterborough, Doncaster, York, Northallerton, Darlington, Durham and Newcastle.
She notes the following:

- All trains stop at Peterborough, York, Darlington and Newcastle.
- All trains which stop at Northallerton also stop at Durham.
- Each day, 50% of the trains stop at both Newark *and* Northallerton.
- All designated "Fast" trains make less than 5 stops. All other trains make 5 stops or more.
- On average, 16 trains run each day.

Which of the following can be reliably concluded from these observations?

A. All trains, which are not designated "fast" trains, must stop at Durham.
B. No more than 8 trains on any 1 day will stop at Northallerton.
C. No designated "Fast" trains will stop at Durham.
D. It is possible for a train to make 5 stops, including Northallerton.
E. A train which stops at Newark will also stop at Durham.

Question 45:

Rakton is 5 miles directly north of Blueville. Gallford is 8 miles directly south of Haston. Lepstone is situated 5 miles directly east of Blueville, and 5 miles directly west of Gallford.

Which of the following **CANNOT** be reliably concluded from this information?

A. Lepstone is South of Rakton
B. Haston is North of Rakton
C. Gallford is East of Rakton
D. Blueville is East of Haston
E. Haston is North of Lepstone

Question 46:

The Eastminster Parliament is undergoing a new set of elections. There are 600 seats up for election, each of which will be elected separately by the people living in that constituency. 6 parties win at least 1 seat in the election, the Blue Party, the Red party, the Orange party, the Yellow party, the Green party and the Purple party. In order to form a government, a party (or coalition) must hold *over* 50% of the seats. After the election, a political analysis committee produces the following report:

- No party has gained more than 45% of the seats, so nobody is able to form a government by themselves.
- The red and the blue party each gained over 40% of the seats.
- No other party gained more than 4% of the seats.
- The green party gained the 4th highest number of seats.

The red party work out that if they collaborate with the green party and the orange party, between the 3 of them, they will have enough seats to form a coalition government.

What is the minimum number of seats that the green party could have?

A. 5
B. 6
C. 7
D. 8
E. 9
F. 10

Questions 47-51 are based on the following information:

A grandmother wants to give her 5 grandchildren £100 between them for Christmas this year. She wants to grade the money she gives to each grandchild exactly so that the older children receive more than the younger ones. She wants share the money such that she will give the 2nd youngest child as much more than the youngest, as the 3rd youngest gets than the 2nd youngest, as the 4th youngest gets from the 3rd youngest and so on. The result will be that the two youngest children together will get seven times as less money than the three oldest.

M is the amount of money the youngest child receives, and D the difference between the amount the youngest and 2nd youngest children receive.

Question 47:

What is the expression for the amount the oldest child receives?

A. M
B. $M + D$
C. $2M$
D. $4M^2$
E. $M + 4D$
F. None of the above.

Question 48:

What is the correct expression for the total money received?

A. $5M = £100$
B. $5D + 10M = £100$
C. $D = \frac{M}{100}$
D. $5M + 10D = £100$
E. $M = \frac{2D}{11}$

Question 49:

"The two youngest children together will get seven times less money than the three oldest."
Which one of the following best expresses the above statement?

A. $7(3M + 9D) = 2M + D$
B. $7D = M$
C. $7(2M + D) = 3M + 9D$
D. $2(7M + D) = 3M + 9D$

Question 50:
Using the statement in the previous question, what is the correct expression for M?

A. $\frac{2D}{11}$ B. $\frac{2}{11}$ C. $\frac{10D}{11}$ D. $\frac{120}{11}$

Question 51:
Express £100 in terms of D.

A. $£\,100 = \dfrac{120D}{11}$

B. $£\,100 = \dfrac{120D}{10}$

C. $£\,100 = \dfrac{120}{11D}$

D. $£\,100 = 21\text{D}$

E. $£\,100 = 5M + 10D$

Question 52:
Four young girls entered a local baking competition. Though a bit burnt, Ellen's carrot cake did not come last. The girl who baked a Madeira sponge had practiced a lot, and so came first, while Jaya came third with her entry. Aleena did better than the girl who made the Tiramisu, and the girl who made the Victoria sponge did better than Veronica.

Which **TWO** of the following were **NOT** results of the competition?
A. Veronica made a tiramisu
B. Ellen came second
C. Aleena made a Victoria sponge
D. The Victoria sponge came in 3rd place
E. The carrot cake came 3rd

Question 53:
In a young children's football league of 5 teams were; Celtic Changers, Eire Lions, Nordic Nesters, Sorten Swipers and the Whistling Winners. One of the boys playing in the league, after being asked by his parents, said that while he could remember the other teams' total points he could not remember his own, the Eire Lions, score. He said that all the teams played each other and when teams lost they were given 0 points, when they drew, 1 point, and 3 for a win. He remembered that the Celtic Changers had a total of 2 points; the Sorten Swipers had 5; the Nordic Nesters had 8, and the Whistling Winners 1.

How many did the boy's team score?

A. 1
B. 4
C. 8
D. 10
E. 11
F. None of the above.

Question 54:
T is the son of Z, Z and J are sisters, R is the mother of J and S is the son of R.

Which one of the following statements is correct?
A. T and J are cousins
B. S and J are sisters
C. J is the maternal uncle of T
D. S is the maternal uncle of T
E. R is the grandmother of Z.

Question 55:
John likes to shoot bottles off a shelf. In the first round he places 16 bottles on the shelf and knocks off 8 bottles. 3 of the knocked off bottles are damaged and can no longer be used, whilst 1 bottle is lost. He puts the undamaged bottles back on the shelf before continuing. In the second round he shoots six times and misses 50% of these shots. He damages two bottles with every shot which does not miss. 2 bottles also fall off the shelf at the end. He puts up 2 new bottles before continuing. In the final round, John misses all his shots and in frustration, knocks over gets angry and knocks over 50% of the remaining bottles.

How many bottles were left on the wall after the final round?

A. 2
B. 3
C. 4

D. 5
E. 6

F. More information needed.

Questions 56 - 62 are based on the information below:

All lines are named after a station they serve, apart from the Oval and Rectangle lines, which are named for their recognisable shapes. Trains run in both directions.

- There are express trains that run from end to end of the St Mark's and Straightly lines in 5 and 6 minutes respectively.
- It takes 2 minutes to change between St Mark's and both Oval and Rectangle lines, 1 minute between Rectangle and Oval.
- It takes 3 minutes to change between the Straightly and all other lines, except with the St Mark's line which only takes 30 seconds
- The Straightly line is a fast line and takes only 2 minutes between stops apart from to and from Keyton, which only takes 1 minute, and to and from Lime St which takes 3 minutes.
- The Oval line is much slower and takes 4 minutes between stops, apart from between Baxton and Marven, and also Archite and West Quays, which takes 5 minutes.
- The Rectangle line a reliable line; never running late but as a consequence is much slower taking 6 minutes between stops.
- The St Mark's line is fast and takes 2 and half minutes between stations.
- If a passenger reaches the end of the line, it takes three minutes to change onto a train travelling back in the opposite direction.

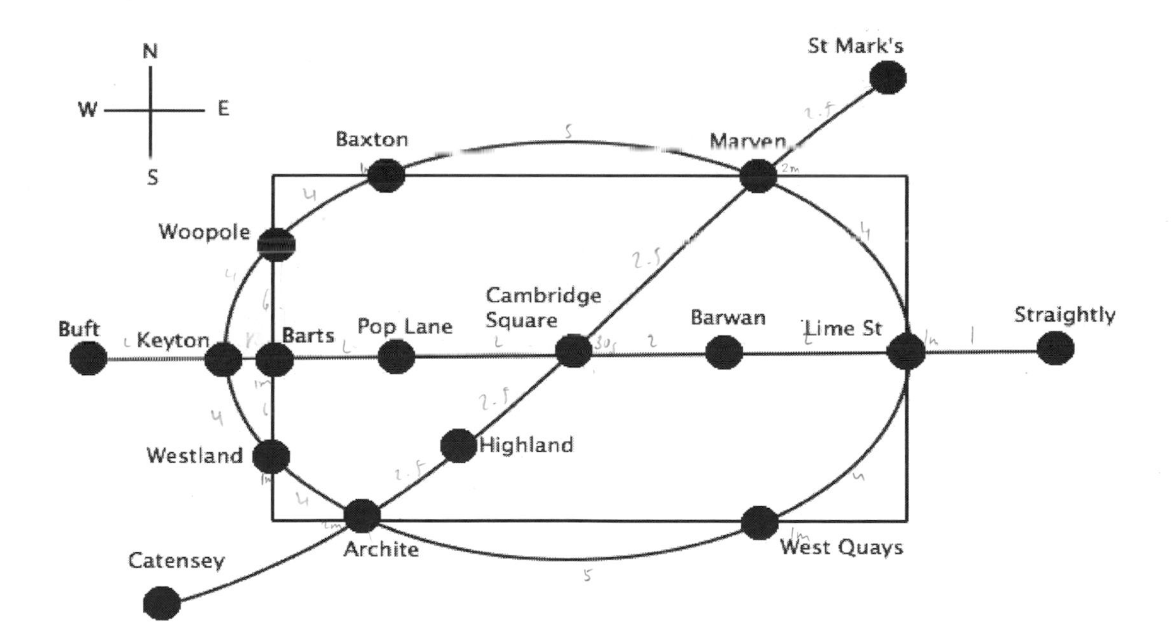

Question 56:

Assuming all lines are running on time, how long does it take to go from St Mark's to Archite on the St Mark's line?

A. 5 minutes
B. 6 minutes
C. 7.5 minutes
D. 10 minutes
E. 12.5 minutes

Question 57:
Assuming all lines are running on time, what's the shortest time it will take to go from Buft to Straightly?

A. 6 minutes
B. 10 minutes
C. 12 minutes
D. 14 minutes
E. 16 minutes

Question 58:
What is the shortest time it will take to go from Baxton to Pop Lane?

A. 11 minutes
B. 12 minutes
C. 13 minutes
D. 14 minutes
E. 15 minutes

Question 59:
Which station, even at the quickest journey time, is furthest in terms of time from Cambridge Square?

A. Catensey
B. Buft
C. Woopole
D. Westland

Questions 60-62 use this additional information:
On a difficult day there are signal problems whereby all lines except the reliable line are delayed, such that train travel times between stations are doubled. These delays have caused overcrowding at the platforms which means that while changeover times between lines are still the same, passengers always have to wait an extra 5 minutes on all of the platforms before catching the next train.

Question 60:
At best, how long will it now take to go from Westland to Marven?

A. 25 minutes
B. 29 minutes
C. 30 minutes
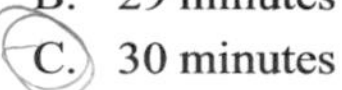
D. 33 minutes
E. 35 minutes

Question 61:
There is a bus that goes from Baxton to Archite and takes 27-31 minutes. Susan lives in Baxton and needs to get to her office in Archite as quickly as possible. With all the delays and lines out of service,

How should you advise Susan best to get to work?

A. Baxton to Archite via Barts using the Rectangle line.
B. Baxton to Woopole on the Rectangle line, then Oval to Archite via Keyton.
C. It is not possible to tell between the fastest two options.
D. Baxton to Woopole on the Rectangle line, then Oval to Archite via Keyton.
E. Baxton to Archite on the Oval line.
F. Baxton to Archite using the bus.

Question 62:
In addition to the delays the Oval line signals fail completely, so the line falls out of service. How long will it now take to go from St Mark's to West Quays as quickly as possible?

A. 35 minutes
B. 30 minutes
C. 33 minutes
D. 29 minutes
E. 30.5 minutes
F. None of the above.

Question 63:

In an unusual horserace, only 4 horses, each with different racing colours and numbers competed. Simon's horse wore number 1. Lila's horse wasn't painted yellow nor blue, and the horse that wore 3, which was wearing red, beat the horse that came in third. Only one horse wore the same number as the position it finished in. Arthur's horse beat Simon's horse, whereas Celia's horse beat the horse that wore number 1. The horse wearing green, Celia's, came second, and the horse wearing blue wore number 4. Which one of the following must be true?

A. Simon's horse was yellow and placed 3rd.
B. Celia's horse was red.
C. Celia's horse was in third place.
D. Arthur's horse was blue.
E. Lila's horse wore number 4.

Question 64:

Jessie plants a tree with a height of 40 cm. The information leaflet states that the plant should grow by 20% each year for the first 2 years, and then 10% each year thereafter.

What is the expected height at 4 years?

A. 58.08 cm
B. 64.89 cm
C. 69.696 cm
D. 89.696 cm
E. 82.944 cm
F. None of the above

Question 65:

A company is required to pay each employee 10% of their wage into a pension fund if their annual total wage bill is above £200,000. However, there is a legal loophole that if the company splits over two sites, the £200,000 bill is per site. The company therefore decides to have an east site, and a west site.

Name	Annual Salary (£)
Luke	47,000
John	78,400
Emma	68,250
Nicola	88,500
Victoria	52,500
Daniel	63,000

Which employees should be grouped at the same site to minimise the cost to the company?

A. John, Nicola, Luke
B. Nicola, Victoria, Daniel
C. Nicola, Daniel, Luke
D. John, Daniel, Emma
E. Luke, Victoria, Emma

Question 66:

A bus takes 24 minutes to travel from White City to Hammersmith with no stops. Each time the bus stops to pick up and/or drop off passengers, it takes approximately 90 seconds. This morning, the bus picked up passengers from 5 stops, and dropped off passengers at 7 stops.

What is the minimum journey time from White City to Hammersmith this morning?

A. 28 minutes
B. 34 minutes
C. 34.5 minutes
D. 36 minutes
E. 37.5 minutes
F. 42 minutes

Question 67:

Sally is making a Sunday roast for her family and is planning her schedule regarding cooking times. The chicken takes 15 minutes to prepare, 75 minutes to cook, and needs to stand for exactly 5 minutes after cooking. The potatoes take 18 minutes to prepare, 5 minutes to boil, then 50 minutes to roast, and must be roasted immediately after boiling, and then served immediately. The vegetables require only 5 minutes preparation time and 8 minutes boiling time before serving, and can be kept warm to be served at any time after cooking. Given that the cooker can only be cooking two items at any given time and Sally can prepare only one item at a time, what should Sally's schedule be if she wishes to serve dinner at 4pm and wants to start cooking each item as late as possible?

A. Chicken 2.25, potatoes 2.47, vegetables 2.42
B. Chicken 2.25, potatoes 2.47, vegetables 3.47
C. Chicken 2.35, potatoes 3.47, vegetables 2.47
D. Chicken 2.35, potatoes 2.47, vegetables 3.47
E. Chicken 2.45, potatoes 3.47, vegetables 2.47
F. Chicken 2.45, potatoes 2.47, vegetables 3.47

Question 68:

The Smiths have 4 children whose total age is 80. Paul is double the age of Jeremy. Annie is exactly half way between the ages of Jeremy and Paul, and Rebecca is 2 years older than Paul. How old are each of the children?

A. Paul 23, Jeremy 12, Rebecca 26, Annie 19.
B. Paul 22, Jeremy, 11, Rebecca 24, Annie 16.
C. Paul 24, Jeremy 12, Rebecca 26, Annie 18.
D. Paul 28, Jeremy 14, Rebecca 30, Annie 21.
E. More information needed

Question 69:

Sarah has a jar of spare buttons that are a mix of colours and sizes. The jar contains the following assortment of buttons:

	10mm	25mm	40mm
Cream	15	22	13
Red	6	15	7
Green	9	19	8
Blue	20	6	15
Yellow	4	8	26
Black	17	16	14
Total	**71**	**86**	**83**

Sarah wants to use a 25mm diameter button, but doesn't mind if it is cream or yellow. What is the maximum number of buttons she will have to remove in order to guarantee to pick a suitable button on the next attempt?

A. 210
B. 218
C. 219
D. 239
E. None of the above

Question 70:

Ben wants to optimise his score with one throw of a dart. 50% of the time he hits a segment to either side of the one he is aiming at. With this in mind, which segment should he aim for?
[Ignore all double/triple modifiers]

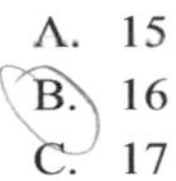

A. 15
B. 16
C. 17
D. 18
E. 19
F. 20

Question 71:

Victoria is completing her weekly shop, and the total cost of the items is £8.65. She looks in her purse and sees that she has a £5 note, and a large amount of change, including all types of coins. She uses the £5 note, and pays the remainder using the maximum number of coins possible in order to remove some weight from the purse. However, the store has certain rules she has to follow when paying:

- No more than 20p can be paid in "bronze" change (the name given to any combination of 1p pieces and 2p pieces)
- No more than 50p can be paid using any combination of 5p pieces and 10p pieces.
- No more than £1.50 can be paid using any combination of 20p pieces and 50p pieces.

Victoria pays the exact amount, and does not receive any change. Under these rules, what is the *maximum* number of coins that Victoria can have paid with?

A. 30
B. 31
C. 36
D. 41
E. 46

Question 72:

I look at the clock on my bedside table, and I see the following digits:

However, I also see that there is a glass of water between me and the clock, which is in front of 2 adjacent figures. I know that this means these 2 figures will appear reversed. For example, 10 would appear as 01, and 20 would appear as 05 (as 5 on a digital clock is a reversed image of a 2). Some numbers, such as 3, cannot appear reversed because there are no numbers which look like the reverse of 3.

Which of the following could be the actual time?

A. 15:52
B. 21:25
C. 12:55
D. 12:22
E. 21:52

Question 73:

Slavica has invaded Worsid, whilst Nordic has invaded Lorkdon. Worsid, spotting an opportunity to bolster its amount of land and natural resources, invades Nordic. Each of these countries is either a dictatorship or a democracy. Slavica is a dictatorship, but Lorkdon is a democracy. 10 years ago, a treaty was signed which guaranteed that no democracy would invade another democracy. No dictatorship has both invaded another dictatorship *and* been invaded by another dictatorship.

Assuming the aforementioned treaty has been upheld, what style of government is practiced in Worsid?

A. Worsid is a Dictatorship.
B. Worsid is a Democracy.

C. Worsid does not practice either of these forms of government.

D. It is impossible to tell.

Question 74:

Sheila is on a shift at the local supermarket. Unfortunately, the till has developed a fault, meaning it cannot tell her how much change to give each customer. A customer is purchasing the following items, at the following costs:

- A packet of grated cheese priced at £3.25
- A whole cucumber, priced at 75p
- A fish pie mix, priced at £4.00
- 3 DVDs, each priced at £3.00

Sheila knows there is an offer on DVDs in the store at present, in which 3 DVDs bought together will only cost £8.00.The customer pays with a £50 note.

How much change will Sheila need to give the customer?

A. £4

B. £33

C. £34

D. £36

E. £38

Question 75:

Ryan is cooking breakfast for several guests at his hotel. He is frying most of the items using the same large frying pan, to get as much food prepared in as little time as possible. Ryan is cooking Bacon, Sausages, and eggs in this pan. He calculates how much room is taken up in the pan by each item. He calculates the following:

- Each rasher of bacon takes up 7% of the available space in the pan
- Each sausage takes up 3% of the available space in the pan.
- Each egg takes up 12% of the available space in the pan.

Ryan is cooking 2 rashers of bacon, 4 sausages and 1 egg for each guest. He decides to cook all the food for each guest at the same time, rather than cooking all of each item at once.

How many guests can he cook for at once?

A. 1

B. 2

C. 3

D. 4

E. 5

Question 76:

SafeEat Inc. is a national food development testing agency. The Manchester-based laboratory has a system for recording all the laboratory employees' birthdays, and presenting them with cake on their birthday, in order to keep staff morale high. Certain amounts of petty cash are set aside each month in order to fund this. 40% of the staff have their birthday in March, and the secretary works out that £60 is required to fund the birthday cake scheme during this month.

If all birthdays cost £2 to provide a cake for, how many people work at the laboratory?

A. 45

B. 60

C. 75

D. 100

E. 150

Question 77:

Many diseases, such as cancer, require specialist treatment, and thus cannot be treated by a general practitioner. Instead, these diseases must be *referred* to a specialist after an initial, more generalised, medical assessment. Bob has had a biopsy on the 1st of August on a lump found in his abdomen. The results show that it is a tumour, with a slight chance of becoming metastatic, so he is referred to a waiting list for specialist radiotherapy and chemotherapy. The average waiting time in the UK for such treatment is 3 weeks, but in Bob's local district, high demand means that it takes 50% longer for each patient to receive treatment. As he is a lower risk case, with a low risk of metastasis, his waiting time is extended by another 20%.

How many weeks will it be before Bob receives specialist treatment?

A. 4.5
B. 4.6
C. 5.0
D. 5.1
E. 5.4
F. 5.6

Question 78:

In a class of 30 seventeen year old students, 40% drink alcohol at least once a month. Of those who drink alcohol at least once a month, 75% drink alcohol at least once a week. 1 in 3 of the students who drink alcohol at least once a week also smoke marijuana. 1 in 3 of the students who drink alcohol less than once a month also smoke marijuana.

How many of the students in total smoke marijuana?

A. 3
B. 4
C. 6
D. 9
E. 10
F. 15

Question 79:

Complete the following sequence of numbers: 1, 4, 10, 22, 46, …

A. 84
B. 92
C. 94
D. 96
E. 100

Question 80:

If the mean of 5 numbers is 7, the median is 8 and the mode is 3, what must the two largest numbers in the set of numbers add up to?

A. 14
B. 21
C. 24
D. 26
E. 35
F. More information needed.

Question 81:

Ahmed buys 1kg bags of potatoes from the supermarket. 1kg bags have to weigh between 900 and 1100 grams. In the first week, there are 10 potatoes in the bag. The next week, there are only 5. Assuming that the potatoes in the bag in week 1 are all the same weight as each other, and the potatoes in the bag in week 2 are all the same weight as each other, what is the maximum possible difference between the heaviest and lightest potato in the two bags?

A. 50g
B. 70g
C. 90g
D. 110g
E. 130g

Question 82:

A football tournament involves a group stage, then a knockout stage. In the group stage, groups of four teams play in a round robin format (i.e. each team plays every other team once) and the team that wins the most matches in each group proceeds through to a knockout stage. In addition, the single best performing second place team across all the groups gains a place in the knockout stage. In the knockout stage, sets of two teams play each other and the one that wins proceeds to the next round until there are two teams left, who play the final.

If we start with 60 teams, how many matches are played altogether?

A. 75
B. 90
C. 100
D. 105
E. 165

Question 83:

The last 4 digits of my card number are 2 times my PIN number, plus 200. The last 4 digits of my husband's card number are the last four digits of my card number doubled, plus 200. My husband's PIN number is 2 times the last 4 digits of his card number, plus 200. Given that all these numbers are 4 digits long, whole numbers, and cannot begin with 0, what is the largest number my PIN number can be?

A. 1,074
B. 1,174
C. 2,348
D. 4,096
E. 9,999
F. More information needed.

Question 84:

All women between 50 and 70 in the UK are invited for breast cancer screening every 3 years. Patients at Doddinghurst Surgery are invited for screening for the first time at any point between their 50th and 53rd birthday. If they ignore an invitation, they are sent reminders every 5 months. We can assume that a woman is screened exactly 1 month after she is sent the invitation or reminder that she accepts. The next invitation for screening is sent exactly 3 years after the previous screening.

If a woman accepts the screening on the second reminder each time, what is the youngest she can be when she has her 4th screening?

A. 60
B. 61
C. 62
D. 63
E. 64
F. 65

Question 85:

Ellie gets a pay rise of k thousand pounds on every anniversary of joining the company, where k is the number of years she has been at the company. She currently earns £40,000, and she has been at the company for 5.5 years. What was her salary when she started at the company?

A. £25,000
B. £27,000
C. £28,000
D. £30,000
E. £31,000
F. £32,000

Question 86:

Northern Line trains arrive into Kings Cross station every 8 minutes, Piccadilly Line trains every 5 minutes and Victoria Line trains every 2 minutes. If trains from all 3 lines arrived into the station exactly 15 minutes ago, how long will it be before they do so again?

A. 24 minutes
B. 25 minutes
C. 40 minutes
D. 60 minutes
E. 65 minutes
F. 80 minutes

Question 87:

If you do not smoke or drink alcohol, your risk of getting Disease X is 1 in 12. If you smoke, you are half as likely to get Disease X as someone who does not smoke. If you drink alcohol, you are twice as likely to get Disease X. A new drug is released that halves anyone's total risk of getting Disease X for each tablet taken. How many tablets of the drug would someone who drinks alcohol have to take to reduce their risk to the same level as someone who smoked but did not take the drug?

A. 0
B. 1
C. 2
D. 3
E. 4
F. 5

Questions 88 – 90 refer to the following information:

There are 20 balls in a bag. 1/2 are red. 1/10 of those that are not red are yellow. The rest are green except 1, which is blue.

Question 88:

If I draw 2 balls from the bag (without replacement), what is the most likely combination to draw?

A. Red and green
B. Red and yellow
C. Red and red
D. Blue and yellow

Question 89:

If I draw 2 balls from the bag (without replacement), what is the least likely (without being impossible) combination to draw?

A. Blue and green
B. Blue and yellow

C. Yellow and yellow
D. Yellow and green

Question 90:
How many balls do you have to draw (without replacement) to guarantee getting at least one of at least three different colours?

A. 5
B. 12
C. 13
D. 17
E. 18
F. 19

Question 91:
A general election in the UK resulted in a hung parliament, with no single party gaining more than 50% of the seats. Thus, the main political parties are engaged in discussion over the formation of a coalition government. The results of this election are shown below:

Political Party	Seats won
Conservatives	260
Labour	270
Liberal Democrats	50
UKIP	35
Green Party	20
Scottish National Party	17
Plaid Cymru	13
Sinn Fein	9
Democratic Unionist Party (DUP)	11
Other	14 (14 other parties won 1 seat each)

There are a total of 699 seats, meaning that in order to form a government, any coalition must have at least 350 seats between them. Several of the party leaders have released statements about who they are and are not willing to form a coalition with, which are summarised as follows:

- The Conservative party and Labour are not willing to take part in a coalition together.
- The Liberal Democrats refuse to take part in any coalition which also involves UKIP.
- The Labour party will only form a coalition with UKIP if the Green party are also part of this coalition.
- The Conservative party are not willing to take part in any coalition with UKIP unless the Liberal Democrats are also involved.

Considering this information, what is the minimum number of parties required to form a coalition government?

A) 2
B) 3
C) 4
D) 5
E) 6

Question 92:
On Tuesday, 360 patients attend appointments at Doddinghurst Surgery. Of the appointments that are booked in, only 90% are attended. Of the appointments that are booked in, 1 in 2 are for male patients, the remaining appointments are for female patients. Male patients are three times as likely to miss their booked appointment as female patients.

How many male patients attend appointments at Doddinghurst Surgery on Tuesday?

A. 30
B. 60
C. 130
D. 150
E. 170

Question 93:

Every A Level student at Greentown Sixth Form studies Maths. Additionally, 60% study Biology, 50% study Economics and 50% study Chemistry. The other subject on offer at Greentown Sixth Form is Physics. Assuming every student studies 3 subjects and that there are 60 students altogether, how many students study Physics?

A. 15
B. 24
C. 30
D. 40
E. 60
F. More information needed

Question 94:

100,000 people are diagnosed with chlamydia each year in the UK. An average of 0.6 sexual partners are informed per diagnosis. Of these, 80% have tests for chlamydia themselves. Half of these tests come back positive.

Assuming that each of the people diagnosed has had an average of 3 sexual partners (none of them share sexual partners or have sex with each other) and that the likelihood of having chlamydia is the same for those partners who are tested and those who are not, how many of the sexual partners who were not tested (whether they were informed or not) have chlamydia?

A. 120,000
B. 126,000
C. 136,000
D. 150,000
E. 240,000
F. 252,000

Question 95:

In how many different positions can you place an additional tile to make a straight line of 3 tiles?

A. 6
B. 7
C. 8
D. 9
E. 10
F. 11
G. 12

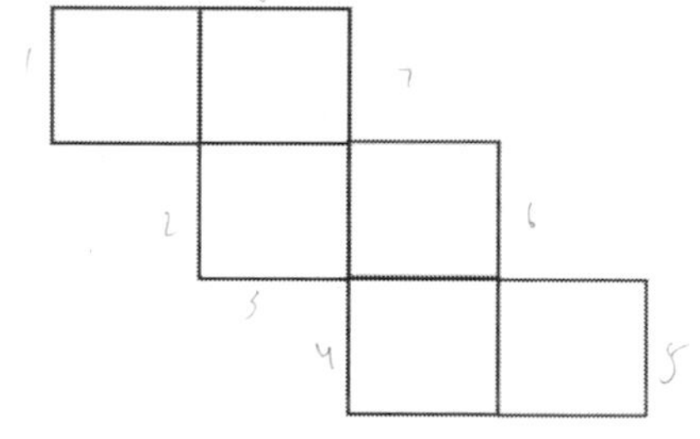

Question 96:

Harry is making orange squash for his daughter's birthday party. He wants to have a 200ml glass of squash for each of the 20 children attending and a 300ml glass of squash for him and each of 3 parents who are helping him out. He has 1,040ml of the concentrated squash.

What ratio of water:concentrated squash should he use in the dilution to ensure he has the right amount to go around?

A. 2:1
B. 3:1
C. 4:1
D. 5:1
E. 6:1
F. 5:2

Question 97:

4 children, Alex, Beth, Cathy and Daniel are each sitting on one of the 4 swings in the park. The swings are in a straight line. One possible arrangement of the children is, left to right, Alex, Beth, Cathy, Daniel.

How many other possible arrangements are there?

A. 5
B. 12
C. 23
D. 24
E. 64
F. 256

Question 98:

A delivery driver is looking to make deliveries in several towns. He is given the following map of the various towns in the area. The lines indicate roads between the towns, along with the lengths of these roads.

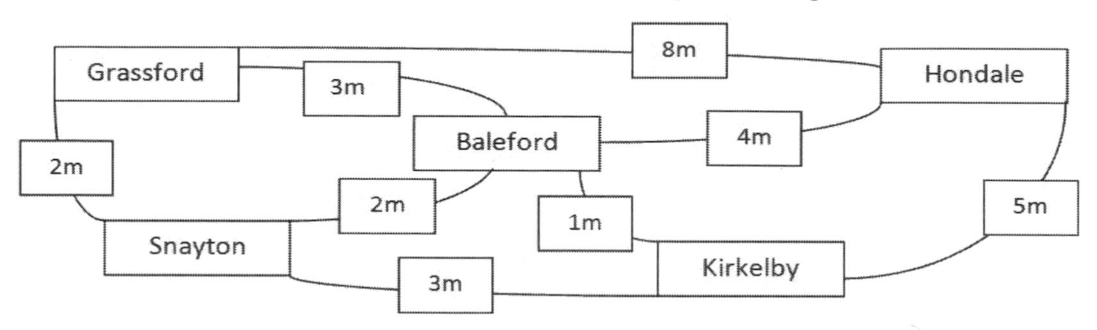

The delivery driver's vehicle has a black box which records the distance travelled and locations visited. At the end of the day, the black box recording shows that he has travelled a total of 14 miles. It also shows that he has visited one town twice, but has not visited any other town more than once. Which of the following is a possible route the driver could have taken?

A. Snayton ? Baleford ? Grassford ? Snayton ? Kirkelby
B. Baleford ? Kirkelby ? Hondale ? Grassford ? Baleford ? Snayton
C. Kirkelby ? Hondale ? Baleford ? Grassford ? Snayton
D. Baleford ? Hondale ? Grassford ? Baleford ? Hondale ? Kirkelby
E. Snayton ? Baleford ? Kirkelby ? Hondale ? Grassford
F. None of the above.

Question 99:

Ellie, her brother Tom, her sister Georgia, her mum and her dad line up in height order from shortest to tallest for a family photograph. Ellie is shorter than her dad but taller than her mum. Georgia is shorter than both her parents. Tom is taller than both his parents.

If 1 is shortest and 5 is tallest, what position is Ellie in the line?

A. 1
B. 2
C. 3
D. 4
E. 5

Question 100:

Miss Briggs is trying to arrange the 5 students in her class into a seating plan. Ashley must sit on the front row because she has poor eyesight. Danielle disrupts anyone she sits next to apart from Caitlin, so she must sit next to Caitlin and no-one else. Bella needs to have a teaching assistant sat next to her. The teaching assistant must be sat on the left hand side of the row, near to the teacher. Emily does not get on with Bella, so they need to be sat apart from one another. The teacher has 2 tables which each sit 3 people, which are arranged 1 behind the other.

Who is sitting in the front right seat?

A. Ashley
B. Bella
C. Caitlin
D. Danielle
E. Emily

Question 101:

My aunt runs the dishwasher twice a week, plus an extra time for each person who is living in the house that week. When her son is away at university, she buys a new pack of dishwasher tablets every 6 weeks, but when her son is home she has to buy a new one every 5 weeks. How many people are living in the house when her son is home?

A. 2
B. 3
C. 4

D. 5
E. 6
F. 7

Question 102:

Dates can be written in an 8 digit form, for example 26-12-2014. How many days after 26-12-2014 would be the next time that the 8 digits were made up of exactly 4 different integers?

A. 6
B. 8
C. 10

D. 16
E. 24
F. 30

Question 103:

Redtown is 4 miles east of Greentown. Bluetown is 5 miles north of Greentown. If every town is due North, South, East or West of at least two other towns, and the only other town is Yellowtown, how many miles away from Yellowtown is Redtown, and in what direction?

A. 4 miles east of Yellowtown.
B. 5 miles south of Yellowtown.
C. 5 miles north of Yellowtown.
D. 4 miles west of Yellowtown.

E. 5 miles west of Yellowtown.
F. None of the above.

Question 104:

Jenna pours wine from two 750ml bottles into glasses. The glasses hold 250ml, but she only fills them to 4/5 of capacity, except the last glass, where she puts whatever she has left. How full is the last glass compared to its capacity?

A. 1/5
B. 2/5

C. 3/5
D. 4/5

E. 5/5

Question 105:

There are 30 children in Miss Ellis's class. Two thirds of the girls in Miss Ellis's class have brown eyes, and two thirds of the class as a whole have brown hair. Given that the class is half boys and half girls, what is the difference between the minimum and maximum number of girls that could have brown eyes and brown hair?

A. 0
B. 2
C. 5

D. 7
E. 10

F. More information needed.

Question 106:

A biased die with the numbers 1 to 6 on it is rolled twice. The resulting numbers are multiplied together, and then their sum subtracted from this result to get the 'score' of the dice roll. If the probability of getting a negative (non-zero) score is 0.75, what is the probability of rolling a 1 on a third throw of the die?

A. 0.1
B. 0.2
C. 0.3

D. 0.4
E. 0.5

F. More information needed.

Questions 107 - 109 are based on the following information:
Fares on the number 11 bus are charged at a number of pence per stop that you travel, plus a flat rate. Emma, who is 21, travels 15 stops and pays £1.70. Charlie, who is 43, travels 8 stops and pays £1.14. Children (under 16) pay half the adult flat rate plus a quarter of the adult charge "per stop".

Question 107:
How much does 17 year old Megan pay to travel 30 stops to college?

A. £0.85
B. £2.40
C. £2.90
D. £3.40
E. More information needed.

Question 108:
How much does 14 year old Alice pay to travel 25 stops to school?

A. £0.50
B. £0.75
C. £1.25
D. £2.50
E. More information needed.

Question 109:
James, who is 24, wants to get the bus into town. The town stop is the 25th stop along a straight road from his house, but he only has £2.

Assuming he has to walk past the stop nearest his house, how many stops will he need to walk past before he gets to the stop he can afford to catch the bus from?

A. 4
B. 6
C. 7
D. 8
E. 9
F. 10

Questions 110 -112 are based on the following information:
Emma mounts and frames paintings. Each painting needs a mount which is 2 inches bigger in each dimension than the painting, and a wooden frame which is 1 inch bigger in each dimension than the mount. Mounts are priced by multiplying 50p by the largest dimension of the mount, so a mount which is 8 inches in one direction and 6 in the other would be £4. Frames are priced by multiplying £2 by the smallest dimension of the frame, so a frame which is 8 inches in one direction and 6 in the other would be £12.

Question 110:
How much would mounting and framing a painting that is 10 x 14 inches cost?

A. £8
B. £26
C. £27
D. £34
E. £42

Question 111:
How much more would mounting and framing a 10 x 10 inch painting cost than mounting and framing an 8 x 8 inch painting?

A. £ 3.00
B. £ 4.00
C. £ 5.00
D. £ 6.00
E. £ 7.00

Question 112:
What is the largest square painting that can be framed for £40?

A. 12 inches
B. 13 inches
C. 14 inches
D. 15 inches
E. 16 inches

Question 113:

If the word 'CREATURES' is coded as 'FTEAWUTEV', which itself would be coded as 'HWEAYUWEX'. What would be the second coding of the word 'MAGICAL'?

A. QCKIGAN
B. OCIIEAN
C. PAJIFAN
D. RALIHAQ
E. RCIMGEP

Question 114:

Jane's mum has asked Jane to go to the shops to get some items that they need. She tells Jane that she will pay her per kilometre that she cycles on her bike to get to the shop, plus a flat rate payment for each place she goes to. Jane receives £6 to go to the grocers, a distance of 5 km, and £4.20 to go the supermarket, a distance of 3km.

How much would she earn if she then cycles to the library to change some books, a distance of 7 km?

A. £7.50
B. £7.70
C. £7.80
D. £8.00
E. £8.10
F. £8.20

Question 115:

In 2001-2002, 1,019 patients were admitted to hospital due to obesity. This figure was more than 11 times higher by 2011-12 when there were 11,736 patients admitted to hospital with the primary reason for admission being obesity.

If the rate of admissions due to obesity continues to increase at the same linear rate as it has from 2001/2 to 2011/12, how many admissions would you expect in 2031/32?

A. 22,453
B. 23,437
C. 33,170
D. 134,964
E. 269,928
F. 300,000

Question 116:

A shop puts its dresses on sale at 20% off the normal selling price. During the sale, the shop makes a 25% profit over the price at which they bought the dresses. What is the percentage profit when the dresses are sold at full price?

A. 36%
B. 42.5%
C. 56.25%
D. 64%
E. 77%
F. 80%

Question 117:

The 'Keys MedSoc committee' is made up of 20 students from each of the 6 years at the university. However, the president and vice-president are sabbatical roles (students take a year out from studying). There must be at least two general committee students from each year, as well as the specialist roles. Additionally, the social and welfare officers must be pre-clinical students (years 1-3) but not first years, and the treasurer must be a clinical student (years 4-6).

Which **TWO** of the following statements must be true?

1. There can be a maximum of 13 preclinical (years 1-3) students on the committee.
2. There must be a minimum of 6 2nd and 3rd years.
3. There is an unequal distribution of committee members over the different year groups.
4. There can be a maximum of 10 clinical (years 4-6) students on the committee.
5. There can be a maximum of 2 first year students on the committee.

6. General committee members are equally spread across the 6 years.

A. 1 and 4
B. 2 and 3
C. 2 and 4
D. 3 and 6
E. 4 and 5
F. 4 and 6

Question 118:

Friday the 13th is superstitiously considered an 'unlucky' day. If 13th January 2012 was a Friday, when would the next Friday the 13th be?

A. March 2012
B. April 2012
C. May 2012
D. June 2012
E. July 2012
F. August 2012
G. September 2012
H. January has the only Friday 13th in 2012.

Question 119:

A farmer has 18 sheep, 8 of which are male. Unfortunately, 9 sheep die, of which 5 were female. The farmer decides to breed his remaining sheep in order to increase the size of his herd. Assuming every female gives birth to two lambs, how many sheep does the farmer have after all the females have given birth once?

A. 10
B. 14
C. 15
D. 16
E. 19

Question 120:

Piyanga writes a coded message for Nishita. Each letter of the original message is coded as a letter a specific number of characters further on in the alphabet (the specific number is the same for all letters). Piyanga's coded message includes the word "PJVN". What could the original word say?

A. CAME
B. DAME
C. FAME
D. GAME
E. LAME

Question 121:

A number of people get on the bus at the station, which is considered the first stop. At each subsequent stop, 1/2 of the people on the bus get off and then 2 people get on. Between the 4th and 5th stop after the station, there are 5 people on the bus.

How many people got on at the station?

A. 4
B. 6
C. 20
D. 24
E. 30

Question 122:

I have recently moved into a new house, and I am looking to repaint my new living room. The price of several different colours of paint is displayed in the table below. A small can contains enough to paint 10 m^2 of wall. A large can contains enough to paint 25 m^2 of wall.

Colour	Cost for a Small Can	Cost for a Large Can
Red	£4	£12
Blue	£8	£15
Black	£3	£9
White	£2	£13

Green	£7	£15
Orange	£5	£20
Yellow	£10	£12

I decide to paint my room a mixture of blue and white, and I purchase some small cans of blue paint and white paint. The cost of blue paint accounts for 50% of the total cost. I paint a total of 100 m^2 of wall space. I use up all the paint. How many m^2 of wall space have I painted blue?

A. 10 m^2
B. 20 m^2
C. 40 m^2
D. 50 m^2
E. 80 m^2

Question 123:

Cakes usually cost 42p at the bakers. The bakers want to introduce a new offer where the amount in pence you pay for each cake is discounted by the square of the number of cakes you buy. For example, buying 3 cakes would mean each cake costs 33p. Isobel says that this is not a good offer from the baker's perspective as it would be cheaper to buy several cakes than just 1. How many cakes would you have to buy for the total cost to fall below 40p?

A. 2
B. 3
C. 4
D. 5
E. 6

Question 124:

The table below shows the percentages of students in two different universities who take various courses. There are 800 students in University A and 1200 students in University B. Biology, Chemistry and Physics are counted as "Sciences".

	University A	University B
Biology	23.50	13.25
Economics	10.25	14.5
Physics	6.25	14.75
Mathematics	11.50	17.25
Chemistry	30.25	7.00
Psychology	18.25	33.25

Assuming each student only takes one course, how many more students in University A than University B study a "Science"?

A. 10
B. 25
C. 60
D. 250
E. 600

Question 125:

Traveleasy Coaches charge passengers at a rate of 50p per mile travelled, plus an additional charge of £5.00 for each international border crossed during the journey. Europremier Coaches charge £15 for every journey, plus 10p per mile travelled, with no charge for crossing international borders. Sonia is travelling from France to Germany, crossing 1 international border. She finds that both companies will charge the same price for this journey.

How many miles is Sonia travelling?

A. 10
B. 20
C. 25
D. 35
E. 40

Question 126:

Lauren, Amy and Chloe live in different cities across England. They decide to meet up together in London and have a meal together. Lauren departs from Southampton at 2:30pm, and arrives in London at 4pm. Amy's journey lasts twice as long as Lauren's journey and she arrives in London at 4:15pm. Chloe departs from Sheffield at 1:30pm, and her journey lasts an hour longer than Lauren's journey.

Which of the following statements is definitely true?

A. Chloe's journey took the longest time.

B. Amy departed after Lauren.
C. Chloe arrived last.
D. Everybody travelled by train.
E. Amy departed before Chloe.

Question 127:

Emma is packing to go on holiday by aeroplane. On the aeroplane, she can take a case of dimension 50cm by 50cm by 20cm, which, when fully packed, can weigh up to 20kg. The empty suitcase weighs 2kg. In her suitcase, she needs to take 3 books, each of which is 0.2m by 0.1m by 0.05m in size, and weighs 1000g. She would also like to take as many items of clothing as possible. Each item of clothing has volume 1500cm³ and weighs 400 g.

Assuming each item of clothing can be squashed so as to fill any shape gap, how many items of clothing can she take in her case?

A. 28
B. 31
C. 34
D. 37
E. 40

Question 128:

Alex is buying a new bed and mattress. There are 5 bed shops Alex can buy the bed and mattress he wants from, each of which sells the bed and mattress for a different price as follows:

- **Bed Shop A:** Bed £120, Mattress £70
- **Bed Shop B:** All beds and mattresses £90 each
- **Bed Shop C:** Bed £140, Mattress £60. Mattress half price when you buy a bed and mattress together.
- **Bed Shop D:** Bed £140, Mattress £100. Get 33% off when you buy a bed and mattress together.
- **Bed Shop E:** Bed £175. All beds come with a free mattress.

Which is the cheapest place for Alex to buy the bed and mattress from?

A. Bed Shop A
B. Bed Shop B
C. Bed Shop C
D. Bed Shop D
E. Bed Shop E

Question 129:

In Joseph's sock drawer, there are 21 socks. 4 are blue, 5 are red, 6 are green and the rest are black. How many socks does he need to take from the drawer in order to guarantee he has a matching pair?

A. 3
B. 4
C. 5
D. 6
E. 7

Question 130:

Printing a magazine uses 1 sheet of card and 25 sheets of paper. It also uses ink. Paper comes in packs of 500 and card comes in packs of 60 which are twice the price of a pack of paper. Each ink cartridge prints 130 sheets of either paper or card. A pack of paper costs £3. Ink cartridges cost £5 each.

How many complete magazines can be printed with a budget of £300?

A. 210
B. 220
C. 230
D. 240
E. 250

Question 131:

Rebecca went swimming yesterday. After a while she had covered one fifth of her intended distance. After swimming six more lengths of the pool, she had covered one quarter of her intended distance. How many lengths of the pool did she intend to complete?

A. 40
B. 72
C. 80
D. 100
E. 120

Question 132:

As a special treat, Sammy is allowed to eat five sweets from his very large jar which contains many sweets of each of three flavours – Lemon, Orange and Strawberry. He wants to eat his five sweets in such a way that no two consecutive sweets have the same flavour.

In how many ways can he do this?

A. 32
B. 48
C. 72
D. 108
E. 162

Question 133:
Granny and her granddaughter Gill both had their birthday yesterday. Today, Granny's age in years is an even number and 15 times that of Gill. In 4 years' time Granny's age in years will be the square of Gill's age in years.

How many years older than Gill is Granny today?

A. 42
B. 49
C. 56
D. 60
E. 64

Question 134:
Pierre said, "Just one of us is telling the truth". Qadr said, "What Pierre says is not true". Ratna said, "What Qadr says is not true". Sven said, "What Ratna says is not true". Tanya said, "What Sven says is not true".

How many of them were telling the truth?

A. 0
B. 1

C. 2
D. 3
E. 4

Question 135:
Two entrants in a school's sponsored run adopt different tactics. Angus walks for half the time and runs for the other half, whilst Bruce walks for half the distance and runs for the other half. Both competitors walk at 3 mph and run at 6 mph. Angus takes 40 minutes to complete the course.

How many minutes does Bruce take?

A. 30
B. 35
C. 40
D. 45
E. 50

Question 136:
Dr Song discovers two new alien life forms on Mars. Species 8472 have one head and two legs. Species 24601 have four legs and one head. Dr Song counts a total of 73 heads and 290 legs in the area. How many members of Species 8472 are present?

A. 0
B. 1
C. 72
D. 73
E. 145
F. More information needed.

Question 137:
A restaurant menu states that:
"All chicken dishes are creamy and all vegetable dishes are spicy. No creamy dishes contain vegetables."

Which of the following **must** be true?

A. Some chicken dishes are spicy.
B. All spicy dishes contain vegetables.
C. Some creamy dishes are spicy.
D. Some vegetable dishes contain tomatoes.
E. None of the above

Question 138:
Simon and his sister Lucy both cycle home from school. One day, Simon is kept back in detention so Lucy sets off for home first. Lucy cycles the 8 miles home at 10 mph. Simon leaves school 20 minutes later than Lucy. How fast must he cycle in order to arrive home at the same time as Lucy?

A. 10 mph
B. 14 mph
C. 17 mph
D. 21 mph
E. 24 mph

Question 139:

Dr. Whu buys 2000 shares in a company at a rate of 50p per share. He then sells the shares for 58p per share. Subsequently he buys 1000 shares at 55p per share then sells them for 61p per share. There is a charge of £20 for each transaction of buying or selling shares. What is Dr. Whu's total profit?

A. £140
B. £160
C. £180
D. £200
E. £220

Question 140:

Jina is playing darts. A dartboard is composed of equal segments, numbered from 1 to 20. She takes three throws, and each of the darts lands in a numbered segment. None land in the centre or in double or triple sections. What is the probability that her total score with the three darts is odd?

A. $1/4$
B. $1/3$
C. $1/2$
D. $3/5$
E. $2/3$

Question 141:

John Morgan invests £5,000 in a savings bond paying 5% interest per annum. What is the value of the investment in 5 years' time?

A. £6,250
B. £6,315
C. £6,381
D. £6,442
E. £6,570

Question 142:

Joe is 12 years younger than Michael. In 5 years the sum of their ages will be 62. How old was Michael two years ago?

A. 20
B. 24
C. 26
D. 30
E. 32

Question 143:

A book has 500 pages. Vicky tears every page out that is a multiple of 3. She then tears out every remaining page that is a multiple of 6. Finally, she tears out half of the remaining pages. If the book measures 15 cm x 30cm and is made from paper of weight 110 gm^{-2}, how much lighter is the book now than at the start?

A. 1,648 g
B. 1,698 g
C. 1,722 g
D. 1,790 g
E. 1,848 g

Question 144:

A farmer is fertilising his crops. The more fertiliser is used, the more the crops grow. Fertiliser costs 80p per kilo. Fertilising at a rate of 0.2 kgm^{-2} increases the crop yield by £1.30 m^{-2}. For each additional 100g of fertiliser above 200g, the extra yield is 30% lower than the linear projection of the stated rate. At what rate of fertiliser application is it no longer cost effective to increase the dose

A. 0.5 kgm^{-2}
B. 0.6 kgm^{-2}
C. 0.7 kgm^{-2}
D. 0.8 kgm^{-2}
E. 0.9 kgm^{-2}

Question 145:

Pet-Star, Furry Friends and Creature Cuddles are three pet shops, which each sell food for various types of pets.

		Price per Kg in:		
Type of pet food	**Amount of food required per week**	**Pet-star**	**Furry Friends**	**Creature Cuddles**
Guinea Pig	3 Kg	£2	£1	£1.50
Cat	6 Kg	£4	£6	£5
Rabbit	4 Kg	£3	£1	£2.50
Dog	8 Kg	£5	£8	£6
Chinchilla	2 Kg	£1.50	£0.50	£1

Given the information above, which of the following statements can we state is definitely *not* true?

A. Regardless of which of these shops you use, the most expensive animal to provide food for will be a dog.
B. If I own a mixture of cats and rabbits, it will be cheaper for me to shop at Pet-star.
C. If I own 3 cats and a dog, the cheapest place for me to shop is at Pet-star
D. Furry Friends sells the cheapest food for the type of pet requiring the most food
E. If I only have one pet, Creature Cuddles will not be the cheapest place to shop regardless of which type of pet I have.

Question 146:

I record my bank balance at the start of each month for six months to help me see how much I am spending each month. My salary is paid on the 10th of each month. At the start of the year, I earn £1000 a month but from March inclusive I receive a pay rise of 10%.

Date	Bank balance
January 1st	1,200
February 1st	1,029
March 1st	1,189
April 1st	1,050
May 1st	925
June 1st	1,025

In which month did I spend the most money?

A. January
B. February
C. March
D. April
E. May

Question 147:

Amy needs to travel from Southtown station to Northtown station, which are 100 miles apart. She can travel by 3 different methods: train, aeroplane or taxi. The tables below show the different times for these 3 methods. The taxi takes 1 minute to cover a distance of 1 mile. Aeroplane passengers must be at the airport 30 minutes before their flight. Southtown airport is 10 minutes travelling time from Southtown station and Northtown airport is 30 minutes travelling time from Northtown station.

If Amy wants to arrive by 1700 and wants to set off as late as possible, what method of travel should she choose and what time will she leave Southtown station?

Train	**Departs Southtown station**	**1400**	**1500**	**1600**
	Arrives Northtown station	1615	1650	1715
Flights	Departs Southtown airport	1610		
	Arrives Northtown airport	1645		

A. Flight, 1530
B. Train, 1600
C. Taxi, 1520
D. Train, 1500
E. Flight, 1610

Question 148:

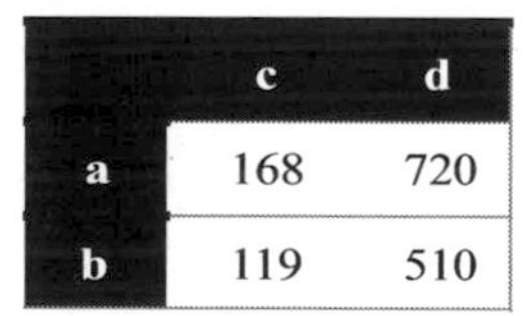

	c	d
a	168	720
b	119	510

In the multiplication grid below, a, b, c and d are all integers. What does d equal?

A. 18
B. 24
C. 30
D. 40
E. 45

Question 149:

A sixth form college has 1,500 students. 48% are girls. 80 of the girls are mixed race.

If an equal proportion of boys and girls are mixed race, how many mixed race boys are there in the college to the nearest 10?

A. 50
B. 60
C. 70
D. 80
E. 90

Question 150:
Christine is a control engineer at the Browdon Nuclear Power Plant. On Wednesday, she is invited to a party on the Friday, and asks her manager if she can take the Friday off. She acknowledged that this will mean she will have worked less than the required number of hours this week, and offers to make this up by working extra hours next week. Her manager suggests that instead, she works 5 hours this Sunday, and 3 extra hours next Thursday to make up the required hours. Christine accepts this proposal. Christine's amended schedule for the week is shown below.

Day	Monday	Tuesday	Wednesday	Thursday	Friday	Saturday	Sunday
Hours worked	8	7	9	6	0	0	5

How many hours was Christine supposed to have worked this week, if she had completed her usual Friday shift?

A. 34
B. 35
C. 36
D. 38
E. 40
F. 42

Question 151:
Leonidas notes that the time on a normal analogue clock is 0340. What is the smaller angle between the hands on the clock?

A. 110°
B. 120°
C. 130°
D. 140°
E. 150°

Question 152:
Sheila is on a shift at the local supermarket. Unfortunately, the till has developed a fault, meaning it cannot tell her how much change to give each customer. A customer is purchasing the following items, at the following costs:

- A packet of grated cheese priced at £3.25
- A whole cucumber, priced at 75p
- A fish pie mix, priced at £4.00
- 3 DVDs, each priced at £3.00

Sheila knows there is an offer on DVDs in the store at present, in which 3 DVDs bought together will only cost £8.00.The customer pays with a £50 note. How much change will Sheila need to give the customer?

A. £33
B. £34
C. £35
D. £36

E. £37

SECTION 1A: Data Analysis

Data analysis questions show a great variation in type and difficulty. The best way to improve with these questions is to do lots of practice questions in order to familiarise yourself with the style of questions.

Options First

Despite the fact that you may have lots of data to contend with, the rule about looking at the options first still stands in this section. This will allow you to register what type of calculation you are required to make and what data you might need to look at for this. Remember, Options ? Question ? Data/Passage.

Working with Numbers

Percentages frequently make an appearance in this section and it's vital that you're able to work comfortably with them. For example, you should be comfortable increasing and decreasing by percentages, and working out inverse percentages too. When dealing with complex percentages, break them down into their components. For example, 17.5% = 10% + 5% + 2.5%.

Graphs and Tables

When you're working with graphs and tables, it's important that you take a few seconds to check the following before actually extracting data from it.

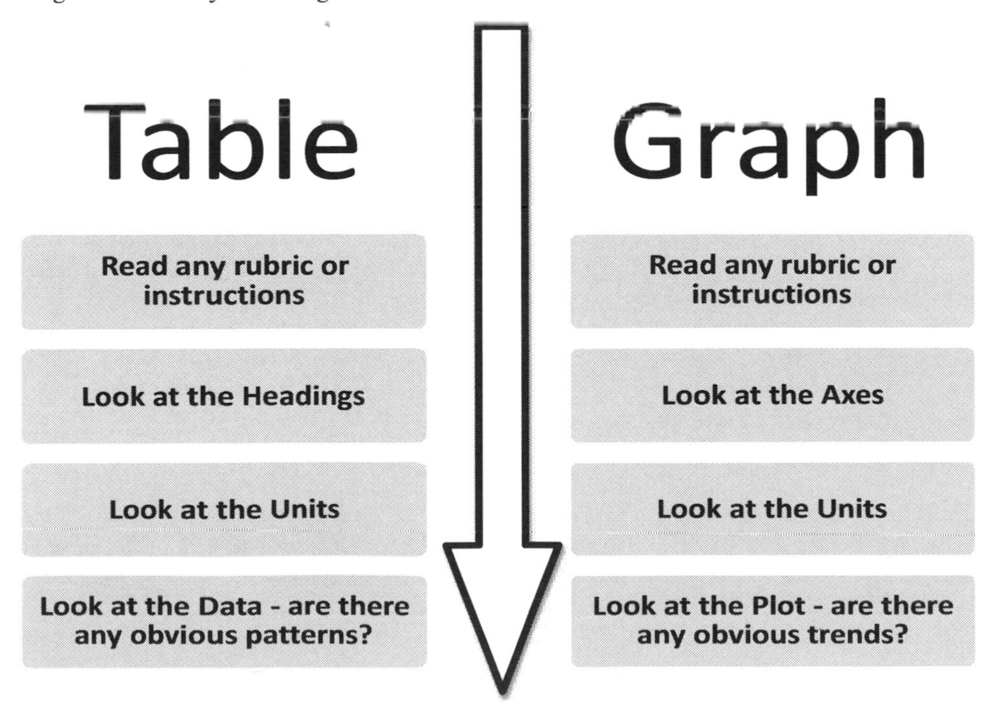

Get into the habit of doing this whenever you are faced with data and you'll find it much easier to approach these questions under time pressure.

SECTION 1A: DATA ANALYSIS

Data Analysis Questions

Questions 153 to 155 are based on the following passage:
It has recently been questioned as to whether the recommended five fruit and vegetables a day is sufficient or if it would be more beneficial to eat 7 fruit and vegetable portions each day. A study at UCL looked at the fruit and vegetables eating habits of 65,000 people in England. Analysis of the data showed that eating more portions was beneficial and vegetables seemed to have a greater protective effect than fruit. The study however did not distinguish whether vegetables themselves have a greater protective effect, or whether these people tend to eat an overall healthier diet. A meta-analysis carried out by researchers across the world complied data from 16 studies which encompassed over 800,000 participants, of whom 56,423 had died.

They found a decline in death of around 5% from all causes for each additional portion of fruit or vegetables eaten, however they recorded no further decline for people who ate over 5 portions. Rates of cardiovascular disease, heart disease or stroke, were shown to decline 4% for each portion up to five, whereas the number of portions of fruit and vegetables eaten seemed to have little impact on cancer rates. The data from these studies points in a similar direction, that eating as much fruit and vegetables a day is preferable, but that five portions is sufficient to have a significant impact on reduction in mortality. Further studies need to look into the slight discrepancies, particularly why the English study found vegetables more protective, and if any specific cancers may be affected by fruit and vegetables even if the general cancer rates more greatly depend on other lifestyle factors.

Question 153:
Which of the following statements is correct?

A. The UCL study found no additional reduction in mortality in those who eat 7 rather than 5 portions of fruit and vegetables a day.
B. People who eat more fruit and vegetables are assumed to have an overall healthier diet which is what gives them the beneficial effect.
C. The meta analysis found fruit and vegetables are more protective against cancer than cardiovascular disease
D. The English study showed fruit had more protective effects than vegetables.
E. The meta-analysis found no additional reduction in mortality in those who eat 7 rather than 5 portions of fruit and vegetables a day.
F. The meta-analysis suggests people who eat 7 portions would have a 10% lower risk of death from any cause than those who eat 5 portions.
G. Fruit and vegetables are not protective against any specific cancers.

Question 154:
If rates of death were found to be 1% lower in the UCL study than the meta-analysis, approximately how many people died in the UCL study?

A. 3,000
B. 3,200
C. 3,900
D. 4,550
E. 5,200

Question 155:
Which statement does the article **MOST** agree with?

A. Eating more fruit and vegetables does not particularly lower the risk of any specific cancers.
B. The UCL research suggests that the guideline should be 7 fruit and vegetables a day for England.
C. The results found by the UCL study and the meta-analysis were contradictory.
D. Many don't eat enough vegetables due to cost and taste.
E. Fruit and vegetables are only protective against cardiovascular disease.
F. The UCL study and meta-analysis use a similar sample of participants.
G. People should aim to eat 7 portions of fruit and vegetables a day.

Questions 156-258 relate to the following table regarding average alcohol consumption in 2010.

Country	Total	Recorded Consumption	Unrecorded consumption	Beer (%)	Wine (%)	Spirits (%)	Other (%)	2020 Consumption Projection
Belarus		14.4	3.2	17.3	5.2	46.6	30.9	17.1
Lithuania	15.4	12.9	2.5		7.8	34.1	11.6	16.2
Andorra	13.8		1.4	34.6		20.1	0	9.1
Grenada	12.5	11.9	0.7	29.3	4.3		0.2	10.4
Czech Republic	13	11.8	1.2	53.5	20.5	26	0	14.1
France	12.2	11.8		18.8	56.4	23.1	1.7	11.6
Russia		11.5	3.6	37.6	11.4	51	0	14.5
Ireland	11.9	11.4	0.5	48.1	26.1	18.7	7.7	10.9

NB: Some data is missing.

Question 156:
Which of the following countries had the highest total beer and wine consumption for 2010?

A. Belarus
B. Lithuania
C. Ireland
D. France
E. Andorra

Question 157:
Which country has the greatest difference for spirit consumption in 2010 and 2020 projection, assuming percentages stay the same?

A. Russia
B. Belarus
C. Lithuania
D. Grenada
E. Ireland

Question 158:
It was later found that some of the percentages of types of alcohol consumed had been mixed up. If the actual amount of beer consumed by each person in the Czech Republic was on average 4.9L, which country were the percentage figures mixed up with?

A. Lithuania
B. Grenada
C. Russia
D. France
E. Ireland
F. Belarus
G. Andorra

Questions 159-162 are based on the following information:

The table below shows the incidence of 6 different types of cancer in Australia:

	Prostate	Lung	Bowel	Bladder	Breast	Uterus
Men	40,000	25,000	20,000	8,000	1,000	0
Women	0	20,000	18,000	4,000	50,000	9,000

Question 159:
Supposing there are 10 million men and 10 million women in Australia, how many percentage points higher is the incidence of cancer amongst women than amongst men?

A. 0.007 %
B. 0.07 %
C. 0.093 %
D. 0.7 %
E. 0.93 %

Question 160:
Now suppose there are 11.5 million men and 10 million women in Australia. Assuming all men are equally likely to get each type of cancer and all women are equally likely to get each type of cancer, how many of the types of cancer are you more likely to develop if you are a man than if you are a woman?

A. 1
B. 2
C. 3
D. 4

Question 161:
Suppose that prostate, bladder and breast cancer patients visit hospital 1 time during the first month of 2015 and patients for all other cancers visit hospital 2 times during the first month of 2015. 10% of cancer patients in Australia are in Sydney, and patients in Sydney are not more or less likely to have certain types of cancer than other patients.

How many hospital visits are made by patients in Sydney with these 6 cancers during the first month of 2015?

A. 10,300
B. 18,400
C. 19,500
D. 28,700
E. 195,000
F. 287,000

Question 162:
Which of the graphs correctly represents the combined proportion of men versus women with bladder cancer?

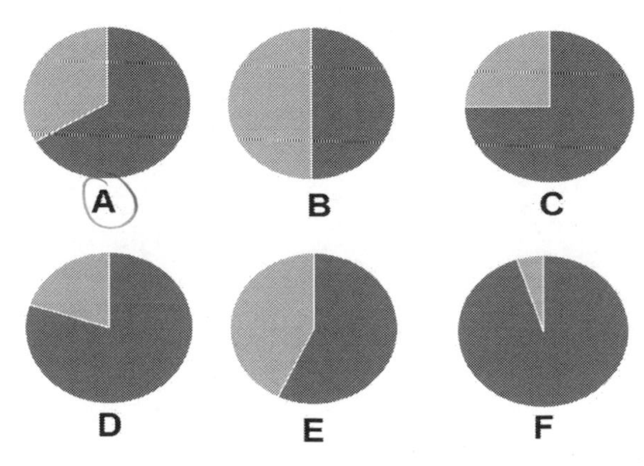

SECTION 1A: DATA ANALYSIS

Questions 163 – 165 are based on the following information:

Units of alcohol are calculated by multiplying the alcohol percentage by the volume of liquid in litres, for example a 0.75 L bottle of wine which is 12% alcohol contains 9 units. 1 pint = 570 ml.

	Volume in bottle/barrel	Standard drinks per bottle/barrel	Percentage
Vodka	1250 ml	50	40%
Beer	10 pints	11.4	3%
Cocktail	750 ml	3	8%
Wine	750 ml	3.75	12.5%

Question 163:

Which standard drink has the most units of alcohol in?

A. Vodka
B. Beer
C. Cocktail
D. Wine

Question 164:

Some guidance suggests the recommended maximum number of units of alcohol per week for women is 14. In a week, Hannah drinks 4 standard drinks of wine, 3 standard drinks of beer, 2 standard cocktails and 5 standard vodkas. This guidance states the recommended maximum number of units per week for men is 21. In a week, Mark drinks 2 standard drinks of wine, 6 standard drinks of beer, 3 standard cocktails and 10 standard vodkas.

Who has exceeded their recommended maximum number of units by more and by how many units more have they exceeded it by than the other person?

A. Hannah, by 1 unit
B. Hannah, by 0.5 units
C. Both by the same
D. Mark, by 0.5 units
E. Mark, by 1 unit

Question 165:

How many different combinations of drinks that total 4 units are there (the same combination in a different order doesn't count).

A. 2
B. 3
C. 4
D. 5
E. 6

Questions 166-168 relate to the table below which shows information about Greentown's population:

	Female	Male	Total
Under 20	1,930		
20-39	1,960	3,760	5,720
40-59		4,130	
60 and over	2,350	2,250	4,600
Total	11,430	12,890	24,320

Question 166:
How many males under 20 are there in Greentown?

A. 2,650
B. 2,700
C. 2,730
D. 2,750
E. 2,850

Question 167:
How many females aged 40-59 are there in Greentown?

A. Between 3,000 and 4,000
B. Between 4,000 and 5,000
C. Between 5,000 and 6,000
D. Between 6,000 and 7,000

Question 168:
Which is the approximate ratio of females:males in the age group that has the highest ratio of males:females?

A. 1.4:1
B. 1.9:1
C. 1:1.9
D. 1:1.4

Questions 169-171 relate to the follow graph:

The graph below shows the average temperatures in London (top trace) and Newcastle (bottom trace).

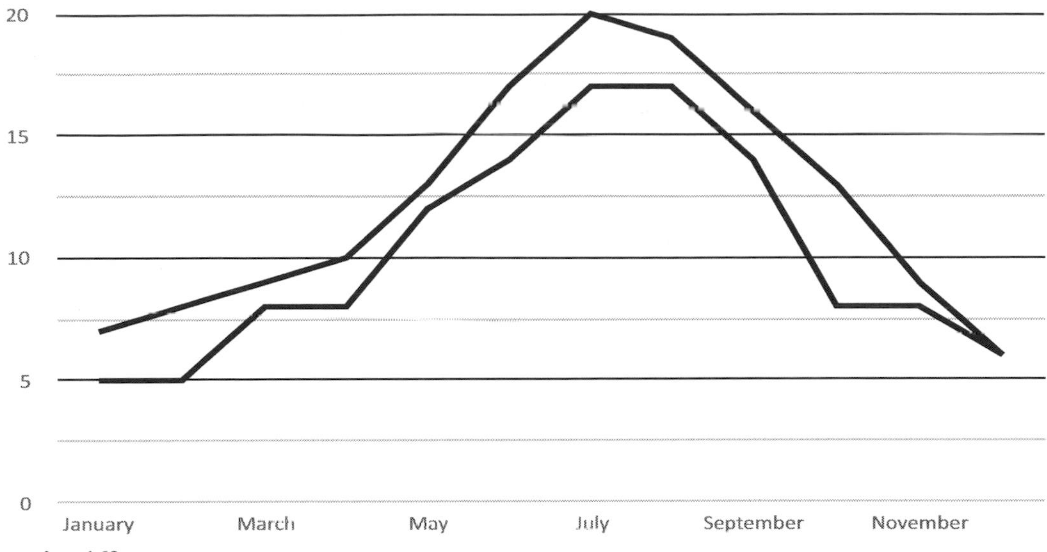

Question 169:

If the average monthly temperature is the same in every year, how many times during the period May 2007 to September 2013 inclusive is the average temperature the same in 2 consecutive months in Newcastle?

A. 20
B. 24
C. 25
D. 30

Question 170:

In how many months in the period specified in the previous question is the average temperature in London AND Newcastle lower than the previous month?

A. 19
B. 21
C. 25
D. 32

Question 171:

To the nearest 0.5 degrees Celcius, what is the average temperature difference between Newcastle and London?

A. 1.5°C
B. 2°C
C. 2.5 °C
D. 3 °C

Questions 172-174 concern the following data:

The pie chart to the right shows sales of ice cream across the four quarters of a year from January to December. Sales are lowest in the month of February. From February they increase in every subsequent month until they get to the maximum sales and from point they decrease in every subsequent month until the of the year.

that end

Question 172:
In which month are the sales highest?

A. June B. July C. August D. Cannot tell

Question 173:
If total sales of ice cream were £354,720 for the year, how much of this was taken during Q1?

A. 29,480
B. £29,560
C. £29,650
D. £29,720
E. £29,800

Question 174:
Assuming total sales revenue (i.e. before costs are taken off) is £180,000, and that each tub of ice cream is sold for £2 and costs the manufacturer £1.50 in total production and transportation costs, how much profit is made during Q2?

A. £15,000
B. £30,000
C. £45,000
D. £60,000

Question 175:
Data on the amount families spend on food per month to the nearest £100 was collected for families with 1, 2 and 3 children. The percentage of families with different spending sizes is displayed below:

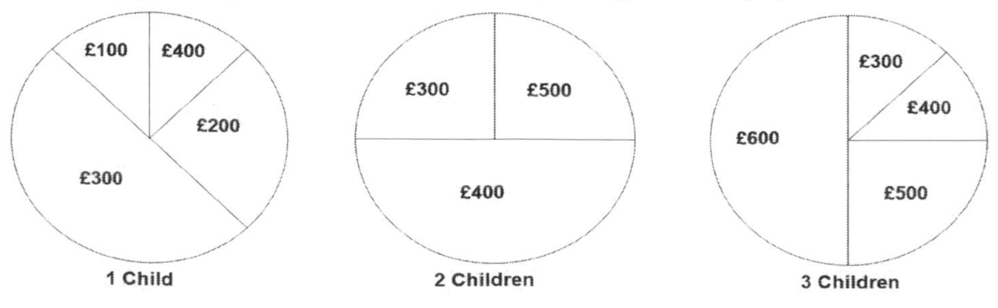

Which of the following statements is definitely true?

A. More families with 1 child than families with 2 children spent £300 a month on food.
B. The overall fraction of families spending £600 was 1/6.
C. All of the families with 2 children spent under £4000 on food per year.
D. The fraction of families with 1 child spending £400 on food per month is the same as the fraction of families with 3 children spending this amount.
E. The average amount spent on food by families with 2 children is £410 a month.

Questions 176-179 are based on the passage below:

A big secondary school recently realised that there were a large number of incidences of bullying occurring that were going unnoticed by teachers. It is possible that some believe bullying to be as much a part of student life as lessons and homework. In order to tackle the problem, the school emailed out a questionnaire to all students' parents and asked them to question their children about where they had experienced or seen bullying in school. Those children that answered yes were then asked if they had told their teachers about it, and asked why they did not if they had not. Those that had told their teacher were asked whether they had seen the teacher act upon the information and whether the bullying had stopped as a result.

Of the 2500 school students surveyed 2210 filled in the online questionnaire. The results were that, 1121 students, almost exactly half (50.7%) had seen bullying in school. Only 396 (35%) of these students told a teacher about the bullying. Of the students who told a teacher, 286 did not witness any action following sharing of the information and of those that did, 60% did not notice any direct action with the bully involved. From those students who did not report the bullying, 146 gave the reason that they didn't think it was important. 427 cited fears of being found out. 212 students said they did not tell because they didn't think the teachers would do anything about it even if they did know. Assume that all the students who filled out the survey did so honestly.

Question 176:
To the nearest integer, what percentage of students did not respond?

A. 10%
B. 12%
C. 18%
D. 8%
E. 5%

Question 177:
If a student saw bullying occur and did not tell a teacher about it, what is the probability that the reasoning for this is that they thought it to be unimportant?

A. 0.1
B. 0.15
C. 0.2
D. 0.35
E. 0.13

Question 178:
After reporting the bullying, how many students saw the teacher act on the information directly with the bully?

A. 66
B. 44
C. 178
D. 104
E. 118

Question 179:
Which of the following does the questionnaire indicate is the best explanation for why students at the school did not report bullying?

A. Students do not think bullying happens at their school.
B. Students think the teachers will do nothing with the information.
C. Students think that bullying is a part of school life.
D. The student's were worried about others finding out.

Question 180:

The obesity epidemic is growing rapidly with reports of a three-fold rise in the period from 2007 to 2012. The rates of hospital admission have also been found to vary massively across different areas of England with the highest rates in the North-East (56 per 100,000 people), and the lowest rates in the East of England (12 per 100,000). During almost every year from 2001-12, there were around twice as many women admitted for obesity as men. The reason for this is however unclear and does not imply there are twice as many obese women as men.

What was the approximate number of admissions per 100 000 women in the North-East in 2011-12?

A. 18
B. 26
C. 37
D. 56
E. 62
F. 74

Question 181:

Health professionals are becoming increasingly worried by the decline in exercise being taken by both children and adults. Around only 40% of adults take the recommended amount of exercise which is 150 minutes per week. As well as falling rates of exercise, a shockingly low number of individuals eat five portions of fruit and vegetables a day. Figures for children aged 5-15 fell to only 16% for boys, and 20% for girls in 2011. Data for adults was only slightly better with 29% of women and 24% of men eating the recommended number of portions.

Using a figure of 8 million children between 5-15 years (equal ratio of girls to boys) in England in 2011, how many more girls than boys ate 5 portions of fruit and vegetables a day?

A. 80,000
B. 120,000
C. 160,000
D. 320,000
E. 640,000

Question 182:

The table below shows the leading causes of death in the UK.

	WOMEN		MEN	
Rank	**Cause of Death**	**Number of Deaths**	**Cause of Death**	**Number of**
1	Dementia and Alzheimer's	31,850	Coronary Heart Disease	37,797
2	Coronary Heart Disease	26,075	Lung Cancer	16,818
3	Stroke	20,706	Dementia and	15,262
4	Flu and Pneumonia	15,361	Lower Respiratory	15,021
5	Lower Respiratory	14,927	Stroke	14,058
6	Lung Cancer	13,619	Flu and Pneumonia	11,426
7	Breast Cancer	10,144	Prostate Cancer	9,726
8	Colon Cancer	6,569	Colon Cancer	7,669
9	Urinary Infections	5,457	Lymphatic Cancer	6,311
10	Heart Failure	5,012	Liver Disease	4,661
	Total	**261,205**	**Total**	**245,585**

Using information from the table only, which of the following statements is correct?

A. More women died from cancers than men.
B. More than 30,000 women died due to respiratory causes.
C. Dementia and Alzheimer's is more common in women than men.

D. No cause of death is of the same ranking for both men and women.

Question 183 is based on the passage below:

The government has recently released a campaign leaflet saying that last year waiting times in NHS A&E departments decreased 20% compared to the year before. The opposition has criticised this statement, saying that there are several definitions which can be described as "waiting times", and the government's campaign leaflet does not make it clear what they mean by "waiting times in A&E".

The NHS watchdog has recently released the following figures describing different aspects of A&E departments, and the change from last year:

Assessment Criterion	2014	2013
Average time spent before being seen in A&E	1 hour	90 minutes
Average time between dialling 999 and receiving treatment in A&E	2 hours	3 hours
Number of people waiting for over 4 hours in A&E	3200	4000
Number of high-priority cases waiting longer than 1 hour	900	1000
Average waiting time for those seen in under 4 hours	50 minutes	40 minutes

Question 183:

Assuming these figures are correct, which criterion of assessment have the government described as "waiting times in A&E" on their campaign leaflet?

A. Number of people waiting for over 4 hours in A&E.
B. Number of people waiting for under 4 hours in A&E.
C. Number of high-priority cases waiting longer than 1 hour.
D. Average time spent before being seen in A&E.
E. Average time between dialling 999 and receiving treatment in A&E.
F. Average waiting time for those seen in less than 4 hours.

Questions 184– 186 refer to the following information:

The table below shows the final standings at the end of the season, after each team has played all the other teams twice each (once at home, once away). The teams are listed in order of how many points they got during the season. Teams get 3 points for a win, 1 point for a draw and 0 points for a loss. No team got the same number of points as another team. Some of the information in the table is missing.

Team	W	D	L
United	8	1	
Athletic	7		
City	7	2	
Town	1	4	
Rovers		0	9
Rangers		2	8

Question 184:
How many points did Rovers get?

A. 0
B. 3
C. 6
D. 9
E. More information needed.

Question 185:
How many games did Athletic lose?

A. 0
B. 1
C. 2
D. 3
E. More information needed.

Question 186:
How many more points did United get than Rangers?

A. 7
B. 15
C. 23
D. 25
E. More information needed.

Questions 187-189 use information from the graph recording A&E attendances and response times for NHS England from 2004 to 2014. Type 1 departments are major A&E units, type 2&3 are urgent care centres or minor injury units. The old target (2004 – June 2010) was 97.5%; the new target (July 2010 – 2015) is 95%.

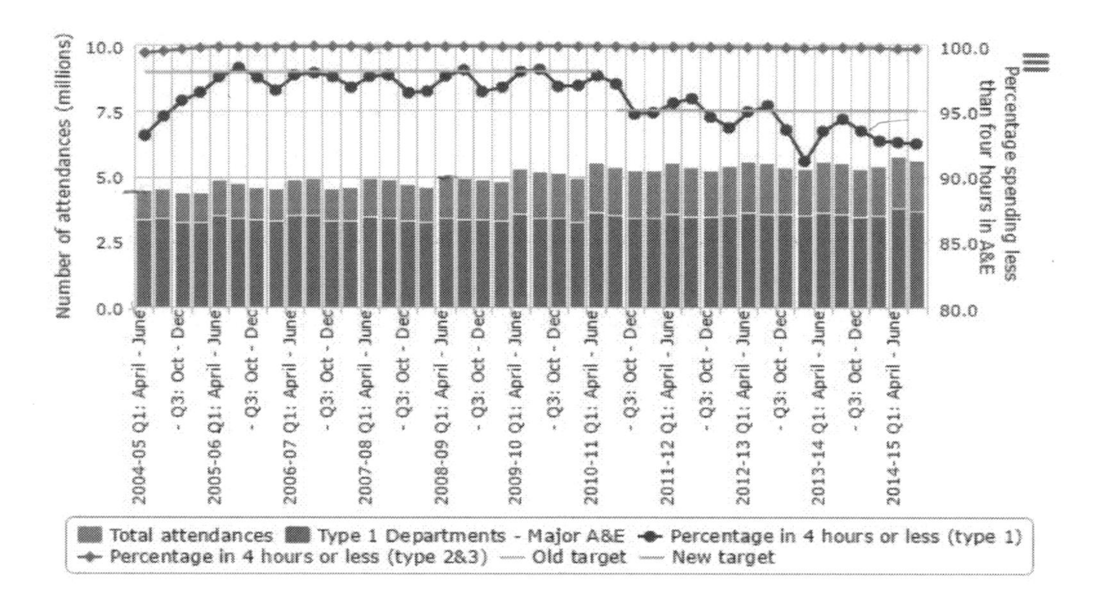

Question 187:

Which of the following statements is **FALSE**?

A. There has been an overall increase in total A&E attendances from 2004-2014.
B. The number of attendances in type 1 departments has been fairly constant from 2004-2014.
C. The new target of 4 hours waiting time has only been reached in two quarters by type 1 departments.
D. The change in attendances is largely due to an in increase people going to type 2&3 departments.

Question 188:
What percentage has the number of total attendances changed from Q1 2004-5 to Q1 2008-9?

A. +5%
B. –5%
C. +10%
D. –10%
E. +15%
F. –15%

Question 189:
If the new target was achieved by type 1 departments 4 times, in what percentage of the quarters was the target missed?

A. 25%
B. 60%
C. 75%
D. 90%

Questions 190-191 relate to the following data:

Ranjna is travelling from Manchester to Bali. She is required to make a stopover in Singapore for which he wants to allow at least 2 hours. It takes 14 hours to fly from Manchester to Singapore, and 2 hours from Singapore to Bali. The table below shows the departure times in local time [Manchester GMT, Singapore GMT + 8, Bali GMT + 8]:

Manchester to Singapore			Singapore to Bali			
Monday	**Wednesday**	**Thursday**	**Monday**	**Tuesday**	**Wednesday**	**Thursday**
08.00	09.30	02.30	13.00	00.00	15.30	13.00
10.45	14.00	08.30	15.30	07.30	18.00	16.00
13.30	18.00	12.30	21.00	08.30	20.30	19.00
15.00	20.00	19.00		12.00		

Question 190:

What is the latest flight Ranjna can take from Manchester to ensure she arrives at Bali Airport by Thursday 22:00?

A. 18:00 Tuesday
B. 14:00 Wednesday
C. 18:00 Wednesday
D. 20:00 Wednesday
E. 02:30 Thursday
F. 08:30 Thursday

Question 191:

Ranjna takes the 08:00 flight from Manchester to Singapore on Monday. She allows 1 hour to clear customs and collect her luggage at Bali Airport and another 45 minutes for the taxi to her hotel. At what time will she arrive at the hotel?

A. 16.45 Monday
B. 04:15 Tuesday
C. 10:30 Tuesday
D. 12:15 Tuesday
E. 12:30 Tuesday
F. 20:30 Tuesday

Question 192:

The graph below represents the percentage of adult smokers in the UK from 1974 to 2010. The top trace represents men and the bottom trace represents women. The middle trace is for both men and women.

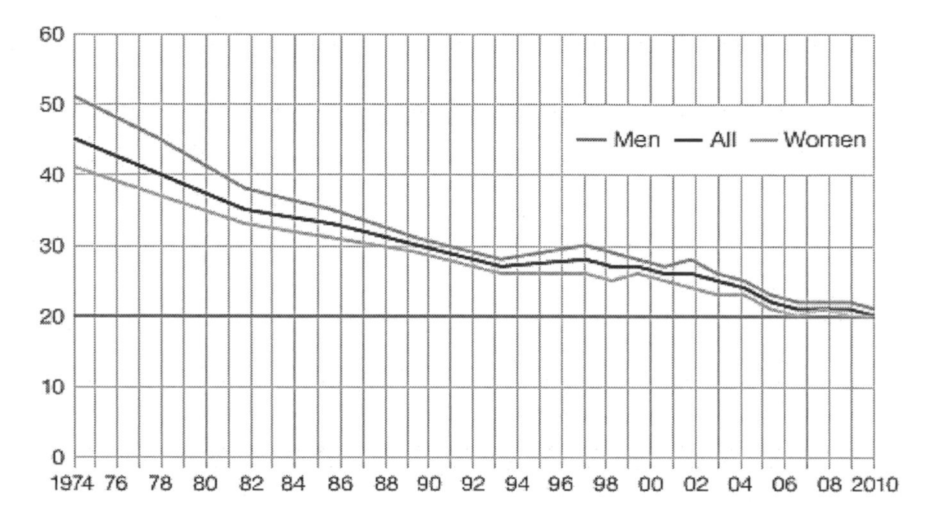

Which of the following statements can be concluded from the graph?

A. The 2007 smoking ban increased the rate in decline of smokers.
B. There has been a constant reduction in percentage of smoker since 1974.
C. The highest rate in decline in smoking for women was 2004-2006.
D. From 1974 to 2010, the smoking rate in men decreased by a half.
E. There has always been a significant difference between the smoking habits of men and women.

Question 193:

The name, age, height, weight and IQ of 11 people were recorded below in a table and a scatter plot. However, the axis labels were left out by mistake. Scale breaks are permitted.

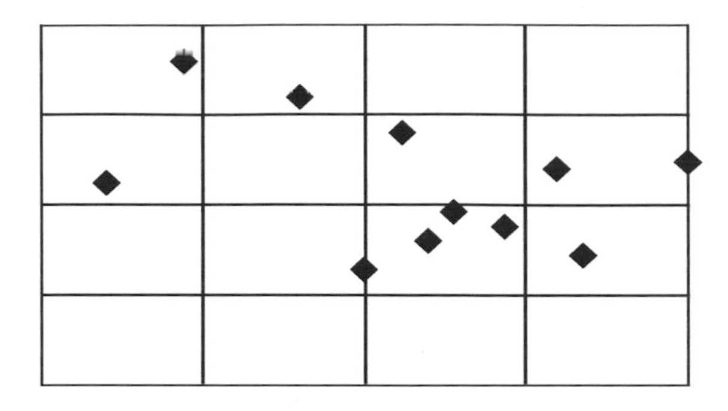

Name	Age	Height (cm)	Weight (kg)	IQ
Alice	18	180	68	110
Ben	12	160	79	120
Camilla	14	170	62	100
David	25	145	98	108
Eliza	29	165	75	96
Rohan	15	190	92	111
George	20	172	88	104
Hannah	22	168	68	115
Ian	13	182	86	98
James	17	176	90	102
Katie	27	151	66	125

Which variants are possible for the X and Y axis?

	X axis	Y axis
A	Height	Weight
B	IQ	Height
C	Age	IQ
D	Height	IQ
E	Height	Age
F	IQ	Weight

Question 194:

A group of students looked at natural variation in height and arm span within their group and got the following results:

Name	Arm span (cm)	Height (cm)
Adam	175	168
Tom	188	175
Shiv	172	184
Mary	148	142
Alice	165	156
Sarah	166	168
Emily	159	160
Matthew	165	172
Michael	185	183

They then drew a scatter plot, but forgot to include names for each point. They also forgot to plot one student.

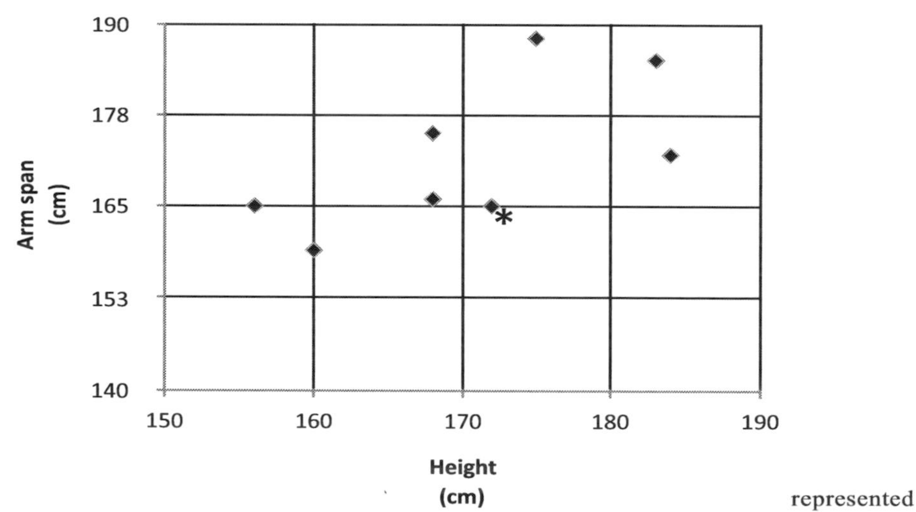

Which student is represented by the point marked with a *?

A. Alice
B. Sarah
C. Matthew
D. Adam
E. Emily
F. Michael

Questions 195 - 201 are based on the following information:
The rectangle represents women. The circle represents those that have children. The triangle represents those that work, and the square those that went to university.

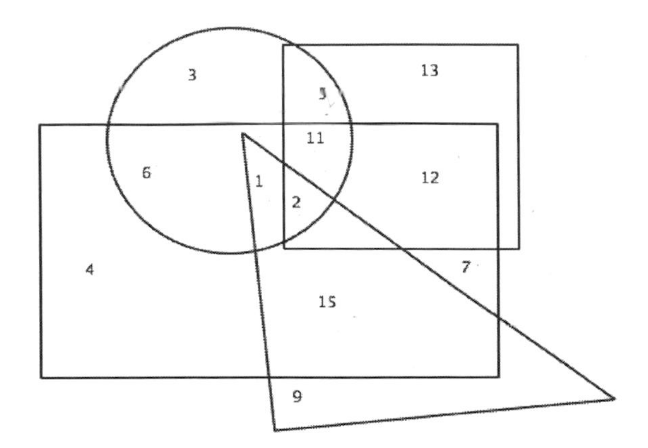

Question 195:
What is the number of non-working women who have children and who did not go to university?

A. 3
B. 5
C. 6
D. 7
E. 9

Question 196:
What is the total number of women who have children and work?

A. 1
B. 2
C. 3
D. 11
E. 14

Question 197:
How many women were surveyed in total?

A. 49
B. 51
C. 58
D. 67
E. 85
F. None of the above.

Question 198:
What is the number of people who went to university and had children?

A. 5
B. 11
C. 13
D. 16
E. 18
F. None of the above.

Question 199:
What is the total number of people who went to university, or have children but not both?

A. 18
B. 28
C. 35
D. 41
E. 53
F. None of the above.

Question 200:
The total number of men who went to university and had children was?

A. 3
B. 4
C. 5
D. 12
E. 13
F. 18

Question 201:
Which of the following people were not surveyed? Choose **TWO** options.

A. A non-working woman who went to university but did not have children.
B. A working man who went to university and has children.
C. A working woman who had children but did not go to university.
D. A non-working man who did not have children and did not go to university.
E. A working woman who went to university but did not have children.

Question 202:
Savers"R"Us is national chain of supermarkets. The price of several items in the supermarket is displayed below:

Item	Price
Beef roasting joint	£8.00
Chicken breast fillet	£6.00
Lamb shoulder	£7.00
Pork belly meat portion	£4.00
Sausages – 10 pack	£3.50

This week the supermarket has a sale on, with 50% off the normal price of all meat products. Alfred visits the supermarket during this sale and purchases a beef roasting joint, a 10 pack of sausages and a lamb shoulder, paying with a £20 note.

How much change does Alfred get?

A. £1.50
B. £5.00
C. £10.75
D. £11.75
E. £12.50
F. None of the above.

Question 203:
The local football league table is shown below, but the number of goals scored against Wilmslow is missing. Each team played the other teams in the league once at home and once away during the season.

Team Name	Points	Goals For	Goals Against
Sale	20	16	2
Wilmslow	16	11	?
Timperley	14	8	7
Altrincham	13	7	9
Mobberley	10	8	12
Hale	8	4	14

How many goals must Wilmslow have conceded?

A. 8
B. 9
C. 10
D. 11
E. 12
F. 14

Weight (lbs)

Height (cm)	100	105	110	115	120
152	19	20	22	24	26
154	18	19	21	23	25
156	17	18	20	22	24
158	15	17	19	21	23
160	14	15	18	20	22
162	13	14	17	19	21
164	12	13	15	18	20
166	11	12	14	17	19
168	10	11	13	15	18
170	9	10	12	14	17

Question 204:

The heights and weights of three women with BMI's 21, 22 and 23 were measured. If Julie and Lydia had different weights but the same height of 154 cm, and the weight of Emma, Lydia and Julie combined was 345 lbs, what was Emma's height?

A. 158 cm
B. 162 cm
C. 160 cm
D. 164 cm
E. 165 cm

	Length (cm)	Weight (lbs)
Bluecup	78	40
Silverfinn	96	60
Starbug	98	98
Jawless	100	56
Lamprene	108	92
Scarfynne	118	40
Rayfish	122	136
Lobefin	126	108
Eringill	146	124
Whaler	148	154
Magic fish	176	124
Blondeye	188	72

Question 205:

The measurements for different types of fish appear below:

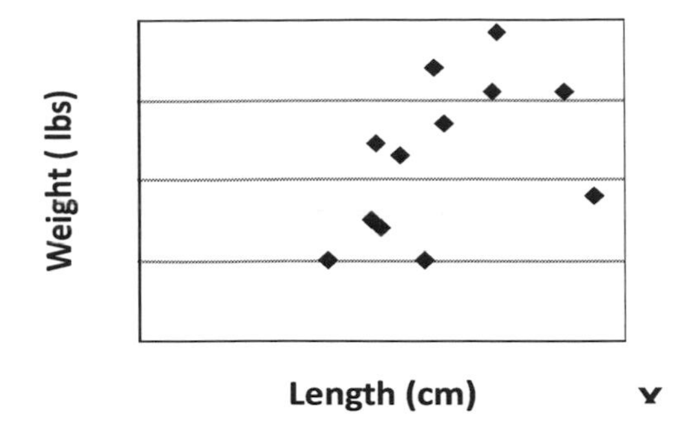

Which fish is shown by the point marked **X**?

A. Silverfinn
B. Starbug
C. Lobefin
D. Blondeye
E. Eringill

The following graphs are required for questions 206-207:
The graph below shows the price of crude oil in US Dollars during 2014:

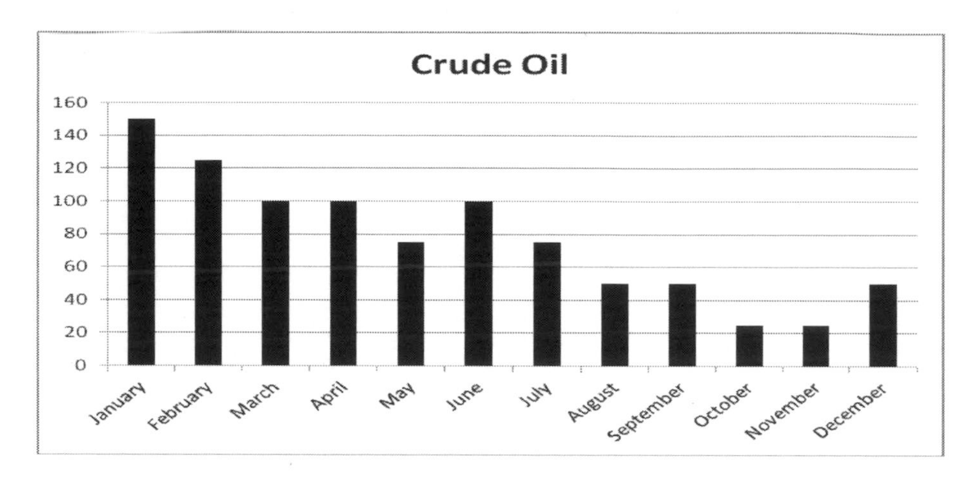

The graph below shows total oil production, in millions of barrels per day:

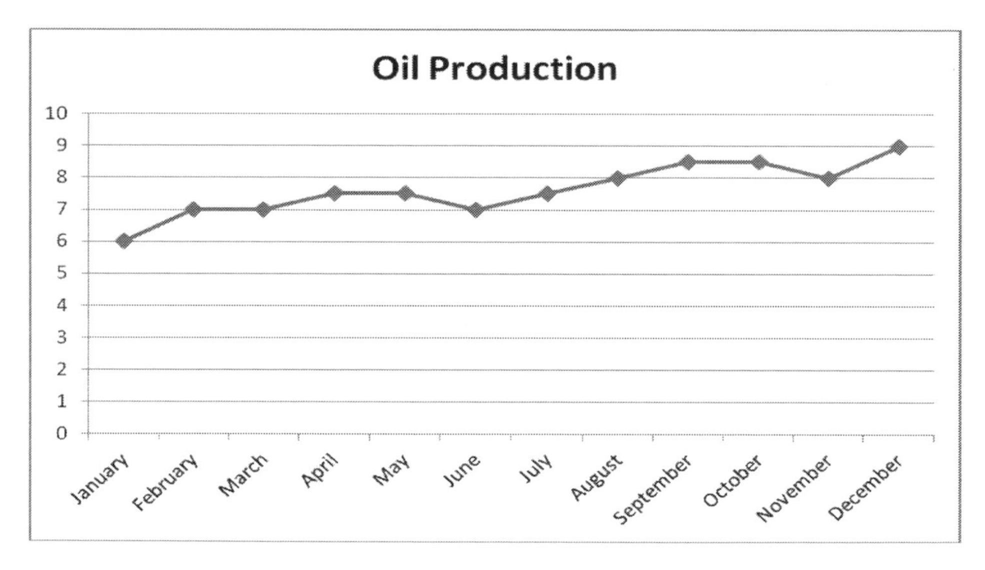

Question 206:
What was approximate total oil production in 2014?

A. 1,750 million barrels
B. 2,146 million barrels
C. 2,300 million barrels
D. 2,700 million barrels
E. 3,500 million barrels

Question 207:
How much did oil sales total in July 2014?

A. $0.56 Billion
B. $16.9 Billion
C. $17.4 Billion
D. $21.1 Billion

SECTION 1B: Advanced Maths

Section 1B tests principles of advanced mathematics. You have to answer 15 questions in 40 minutes. The questions can be quite difficult and it's easy to get bogged down. However, it's well worth spending time preparing for this section as its possible to rapidly improve with targeted preparation.

Gaps in Knowledge

You are highly advised to go through the ECAA Specification and ensure that you have covered all examinable topics. An electronic copy of this can be obtained from **uniadmissions.co.uk/ecaa.** The questions in this book will help highlight any particular areas of weakness or gaps in your knowledge that you may have. Upon discovering these, make sure you take some time to revise these topics before carrying on – there is little to be gained by attempting these questions with huge gaps in your knowledge. A summary of the major topics is given below:

Algebra:

- Laws of Indices
- Manipulation of Surds
- Quadratic Functions: Graphs, use of discrimiant, completing the square
- Solving Simulatenous Equations via Substitution
- Solving Linear and Quadratic Inequalities
- Manipulation of polynomials e.g. expanding brackets, factorising
- Use of Factor Theorem + Remainder Theorem

Graphing Functions:

- Sketching of common functions including lines, quadratics, cubics, trigonometric functions, logarithmic functions and exponential functions
- Manipulation of functions using simple transformaions

Exponentials & Logs:

- Graph of $y = a^x$ series
- Law of Lograithms:
  - $a^b = c \leftrightarrow b = log_a c$
  - $log_a x + log_a y = log_a(xy)$
  - $log_a x - log_a y = log_a(\frac{x}{y})$
  - $k\, log_a x = log_a(x^k)$
  - $log_a \frac{1}{x} = -\, log_a x$

- $$log_a a = 1$$

Trignometry:

- Sine and Cosine rules
- Solution of trigonometric identities
- Values of sin, cost, tan for 0, 30, 45, 60 and 90 degrees
- Sine, Cosine, Tangent graphs, symmetries, perioditicties
- $$Area\ of\ Triangle = \frac{1}{2} ab \sin C$$
- $$\sin^2\theta + \cos^2 = 1$$
- $$tan\theta = \frac{sin\theta}{\cos\theta}$$

Differentiation:

- First order and second order derivatives
- Familiarity with notation: $\frac{dy}{dx}, \frac{d^2y}{dx^2}, f'(x), f''(x)$
- Differentiation of functions like $y = x^n$

Integration:

- Definite and indefinite integrals for $y = x^n$
- Solving Differential Equations in the form: $\frac{dy}{dx} = f(x)$
- Understanding of the Fundamental Theorem of Calculus and its application:
  - $$\int_a^b f(x)dx = F(b) - F(a),\ where\ F'(x) = f(x)$$
  - $$\frac{d}{dx}\int_a^x f(t)dt = f(x)$$

Geometry:

- Circle Properties:
  - The angle subtended by an arc at the centre of a circle is double the size of the angle subtended by the arc on the circumference
  - The opposite angles in a cyclic quadrilateral summate to 180 degrees
  - The angle between the tanent and chord at the point of contact is equal to the angle in the alternate segment
  - The tangent at any point on a circle is perpendicular to the radius at that point

  - Triangles formed using the full diameter are right-angled triangles
  - Angles in the same segment are equal
  - The Perpendicular from the centre to a chord bisects the chord

➢ Equations for a circle:

  - $$(x - a)^2 + (y - b)^2 = r^2$$
  - $$x^2 + y^2 + cx + dy + e = 0$$

➢ Equations for a straight line:

  - $$y - y_1 = m(x - x_1)$$
  - $Ax + by + c = 0$

Series:

➢ Arithmetic series and Geometric Series

➢ Summing to a finite and infinite geometric series

➢ Binomial Expansions

➢ Factorials

Logic Arguments:

➢ Terminology: True, flase, and, or not, necessary, sufficient, for all, for some, there exists.

➢ Arguments in the format:

  - If A then B
  - A if B
  - A only if B
  - A if and only if B

Formulas you **MUST** know:

2D Shapes		**3D Shapes**		
	Area		**Surface Area**	**Volume**
Circle	πr^2	**Cuboid**	Sum of all 6 faces	Length x width x height
Parallelogram	Base x Vertical height	**Cylinder**	$2\pi r^2 + 2\pi rl$	$\pi r^2 \times l$
Trapezium	0.5 x h x (a+b)	**Cone**	$\pi r^2 + \pi rl$	$\pi r^2 \times (h/3)$
Triangle	0.5 x base x height	**Sphere**	$4\pi r^2$	$(4/3)\pi r^3$

Even good students who are studying maths at A level can struggle with certain ECAA maths topics because they're usually glossed over at school. These include:

Quadratic Formula

The solutions for a quadratic equation in the form $ax^2 + bx + c = 0$ are given by: $x = \frac{-b \pm \sqrt{b^2 - 4ac}}{2a}$

Remember that you can also use the discriminant to quickly see if a quadratic equation has any solutions:

$$If\, b^2 - 4ac < 0: No\ solutions$$
$$If\, b^2 - 4ac = 0: One\ solution$$
$$If\, b^2 - 4ac > 2: Two\ solutions$$

Completing the Square

If a quadratic equation cannot be factorised easily and is in the format $ax^2 + bx + c = 0$ then you can rearrange it into the form $a\left(x + \frac{b}{2a}\right)^2 + [c - \frac{b^2}{4a}] = 0$

This looks more complicated than it is – remember that in the ECAA, you're extremely unlikely to get quadratic equations where $a > 1$ and the equation doesn't have any easy factors. This gives you an easier equation: $\left(x + \frac{b}{2}\right)^2 + \left[c - \frac{b^2}{4}\right] = 0$ and is best understood with an example.

Consider: $x^2 + 6x + 10 = 0$

This equation cannot be factorised easily but note that: $x^2 + 6x - 10 = (x + 3)^2 - 19 = 0$

Therefore, $x = -3 \pm \sqrt{19}$. Completing the square is an important skill – make sure you're comfortable with it.

Difference between 2 Squares

If you are asked to simplify expressions and find that there are no common factors but it involves square numbers – you might be able to factorise by using the 'difference between two squares'.

For example, $x^2 - 25$ can also be expressed as $(x + 5)(x - 5)$.

SECTION 1B: MATHS

Maths Questions

Questions 208 – 282 are easier than you will get in the exam but have been included for practice.

Question 208:

Robert has a box of building blocks. The box contains 8 yellow blocks and 12 red blocks. He picks three blocks from the box and stacks them up high. Calculate the probability that he stacks two red building blocks and one yellow building block, in **any** order.

A. $\frac{8}{20}$ B. $\frac{44}{95}$ C. $\frac{11}{18}$ D. $\frac{8}{19}$ E. $\frac{12}{20}$ F. $\frac{35}{60}$

Question 209:

Solve $\frac{3x+5}{5} + \frac{2x-2}{3} = 18$

A. 12.11
B. 13.49
C. 13.95
D. 14.2
E. 19
F. 265

Question 210:

Solve $3x^2 + 11x - 20 = 0$

A. 0.75 and $\frac{4}{3}$
B. -0.75 and $\frac{4}{3}$
C. -5 and $\frac{4}{3}$
D. 5 and $\frac{4}{3}$
E. 12 only
F. -12 only

Question 211:

Express $\frac{5}{x+2} + \frac{3}{x-4}$ as a single fraction.

A. $\frac{15x-120}{(x+2)(x-4)}$
B. $\frac{8x-26}{(x+2)(x-4)}$
C. $\frac{8x-14}{(x+2)(x-4)}$
D. $\frac{15}{8x}$
E. 24
F. $\frac{8x-14}{x^2-8}$

Question 212:

The value of p is directly proportional to the cube root of q. When p = 12, q = 27. Find q when p = 24.

A. 32
B. 64
C. 124
D. 128
E. 216
F. 1728

Question 213:

Write 72^2 as a product of its prime factors.

A. $2^6 \times 3^4$

B. $2^6 \times 3^5$

C. $2^4 \times 3^4$

D. 2×3^3

E. $2^6 \times 3$

F. $2^3 \times 3^2$

Question 214:

Calculate: $\frac{2.302 \times 10^5 + 2.302 \times 10^2}{1.151 \times 10^{10}}$

A. 0.0000202

B. 0.00020002

C. 0.00002002

D. 0.00000002

E. 0.000002002

F. 0.000002002

Question 215:

Given that $y^2 + \mathbf{a}y + \mathbf{b} = (y + 2)^2 - 5$, find the values of **a** and **b**.

	a	b
A	-1	4
B	1	9
C	-1	-9
D	-9	1
E	4	-1
F	4	1

Question 216:

Express $\frac{4}{5} + \frac{m - 2n}{m + 4n}$ as a single fraction in its simplest form:

A. $\frac{6m + 6n}{5(m + 4n)}$

B. $\frac{9m + 26n}{5(m + 4n)}$

C. $\frac{20m + 6n}{5(m + 4n)}$

D. $\frac{3m + 9n}{5(m + 4n)}$

E. $\frac{3(3m + 2n)}{5(m + 4n)}$

F. $\frac{6m + 6n}{3(m + 4n)}$

Question 217:

A is inversely proportional to the square root of B. When A = 4, B = 25.

Calculate the value of A when B = 16.

A. 0.8

B. 4

C. 5

D. 6

E. 10

F. 20

Question 218:

S, T, U and V are points on the circumference of a circle, and O is the centre of the circle.

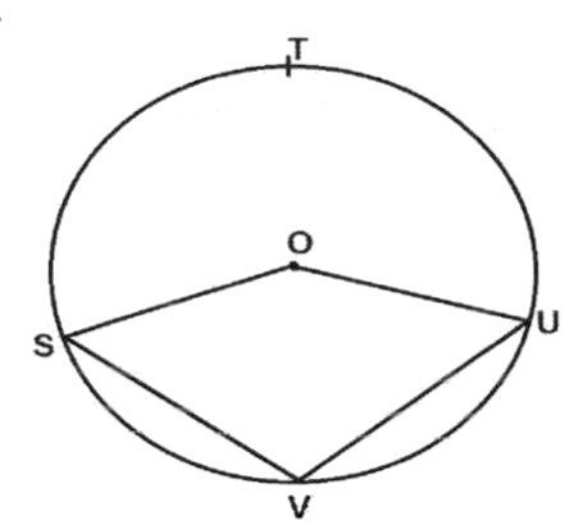

Given that angle SVU = 89°, calculate the size of the smaller angle SOU.

A. 89°
B. 91°
C. 102°
D. 178°
E. 182°
F. 212°

Question 219:

Open cylinder A has a surface area of 8π cm² and a volume of 2π cm³. Open cylinder B is an enlargement of A and has a surface area of 32π cm². Calculate the volume of cylinder B.

A. 2π cm³
B. 8π cm³
C. 10π cm³
D. 14π cm³
E. 16π cm³
F. 32π cm³

Question 220:

Express $\frac{8}{x(3-x)} - \frac{6}{x}$ in its simplest form.

A. $\frac{3x-10}{x(3-x)}$
B. $\frac{3x+10}{x(3-x)}$
C. $\frac{6x-10}{x(3-2x)}$
D. $\frac{6x-10}{x(3+2x)}$
E. $\frac{6x-10}{x(3-x)}$
F. $\frac{6x+10}{x(3-x)}$

Question 221:

A bag contains 10 balls. 9 of those are white and 1 is black. What is the probability that the black ball is drawn in the tenth and final draw if the drawn balls are not replaced?

A. 0
B. $\frac{1}{10}$
C. $\frac{1}{100}$
D. $\frac{1}{10^{10}}$
E. $\frac{1}{362{,}880}$

Question 222:

Gambit has an ordinary deck of 52 cards. What is the probability of Gambit drawing 2 Kings (without replacement)?

A. 0
B. $\frac{1}{169}$
C. $\frac{1}{221}$
D. $\frac{4}{663}$
E. None of the above

Question 223:

I have two identical unfair dice, where the probability that the dice get a 6 is twice as high as the probability of any other outcome, which are all equally likely. What is the probability that when I roll both dice the total will be 12?

A. 0
B. $\frac{4}{49}$
C. $\frac{1}{9}$
D. $\frac{2}{7}$
E. None of the above

Question 224:

A roulette wheel consists of 36 numbered spots and 1 zero spot (i.e. 37 spots in total). What is the probability that the ball will stop in a spot either divisible by 3 or 2?

A. 0

B. $\frac{25}{37}$

C. $\frac{25}{36}$

D. $\frac{18}{37}$

E. $\frac{24}{37}$

Question 225:

I have a fair coin that I flip 4 times. What is the probability I get 2 heads and 2 tails?

A. $\frac{1}{16}$

B. $\frac{3}{16}$

C. $\frac{3}{8}$

D. $\frac{9}{16}$

E. None of the above

Question 226:

Shivun rolls two fair dice. What is the probability that he gets a total of 5, 6 or 7?

A. $\frac{9}{36}$

B. $\frac{7}{12}$

C. $\frac{1}{6}$

D. $\frac{5}{12}$

E. None of the above

Question 227:

Dr Savary has a bag that contains x red balls, y blue balls and z green balls (and no others). He pulls out a ball, replaces it, and then pulls out another. What is the probability that he picks one red ball and one green ball?

A. $\frac{2(x+y)}{x+y+z}$

B. $\frac{xz}{(x+y+z)^2}$

C. $\frac{2xz}{(x+y+z)^2}$

D. $\frac{(x+z)}{(x+y+z)^2}$

E. $\frac{4xz}{(x+y+z)^4}$

F. More information needed

Question 228:

Mr Kilbane has a bag that contains x red balls, y blue balls and z green balls (and no others). He pulls out a ball, does **NOT** replace it, and then pulls out another. What is the probability that he picks one red ball and one blue ball?

A. $\frac{2xy}{(x+y+z)^2}$

B. $\frac{2xy}{(x+y+z)(x+y+z-1)}$

C. $\frac{2xy}{(x+y+z)^2}$

D. $\frac{xy}{(x+y+z)(x+y+z-1)}$

E. $\frac{4xy}{(x+y+z-1)^2}$

F. More information needed

SECTION 1B: MATHS

Question 229:

There are two tennis players. The first player wins the point with probability p, and the second player wins the point with probability 1-p. The rules of tennis say that the first player to score four points wins the game, unless the score is 4-3. At this point the first player to get two points ahead wins.

What is the probability that the first player wins in exactly 5 rounds?

A. $4p^4(1-p)$
B. $p^4(1-p)$
C. $4p(1-p)$
D. $4p(1-p)^4$
E. $4p^5(1-p)$
F. More information needed

Question 230:

Solve the equation $\frac{4x + 7}{2} + 9x + 10 = 7$

A. $\frac{22}{13}$ B. $-\frac{22}{13}$ C. $\frac{10}{13}$ D. $-\frac{10}{13}$ E. $\frac{13}{22}$ F. $-\frac{13}{22}$

Question 231:

The volume of a sphere is $V = \frac{4}{3}\pi r^3$, and the surface area of a sphere is $S = 4\pi r^2$. Express S in terms of V

A. $S = (4\pi)^{2/3}(3V)^{2/3}$
B. $S = (8\pi)^{1/3}(3V)^{2/3}$
C. $S = (4\pi)^{1/3}(9V)^{2/3}$
D. $S = (4\pi)^{1/3}(3V)^{2/3}$
E. $S = (16\pi)^{1/3}(9V)^{2/3}$

Question 232:

Express the volume of a cube, *V*, in terms of its surface area, *S*.

A. $V = (S/6)^{3/2}$
B. $V = S^{3/2}$
C. $V = (6/S)^{3/2}$
D. $V = (S/6)^{1/2}$
E. $V = (S/36)^{1/2}$
F. $V = (S/36)^{3/2}$

Question 233:

Solve the equations $4x + 3y = 7$ and $2x + 8y = 12$

A. $(x, y) = \left(\frac{17}{13}, \frac{10}{13}\right)$
B. $(x, y) = \left(\frac{10}{13}, \frac{17}{13}\right)$
C. $(x, y) = (1, 2)$

D. $(x, y) = (2, 1)$

E. $(x, y) = (6, 3)$

F. $(x, y) = (3, 6)$

G. No solutions possible.

Question 234:

Rearrange $\frac{(7x + 10)}{(9x + 5)} = 3y^2 + 2$, to make x the subject.

A. $\frac{15y^2}{7 - 9(3y^2+2)}$

B. $\frac{15y^2}{7 + 9(3y^2+2)}$

C. $-\frac{15y^2}{7 - 9(3y^2+2)}$

D. $-\frac{15y^2}{7 + 9(3y^2+2)}$

E. $-\frac{5y^2}{7 + 9(3y^2+2)}$

F. $\frac{5y^2}{7 + 9(3y^2+2)}$

Question 235:

Simplify $3x\left(\frac{3x^7}{x^{\frac{1}{3}}}\right)^3$

A. $9x^{20}$
B. $27x^{20}$
C. $87x^{20}$
D. $9x^{21}$
E. $27x^{21}$
F. $81x^{21}$

Question 236:

Simplify $2x[(2x)^7]^{\frac{1}{14}}$

A. $2x\sqrt{2x^4}$

B. $2x\sqrt{2x^3}$

C. $2\sqrt{2x^4}$

D. $2\sqrt{2x^3}$

E. $8x^3$

F. $8x$

Question 237:

What is the circumference of a circle with an area of 10π?

A. $2\pi\sqrt{10}$

B. $\pi\sqrt{10}$

C. 10π

D. 20π

E. $\sqrt{10}$

F. More information needed

Question 238:

If $a \,.\, b = (ab) + (a + b)$, then calculate the value of $(3.4)\,.\,5$

A. 19
B. 54
C. 100
D. 119
E. 132

Question 239:

If $a \cdot b = \frac{a^b}{a}$, calculate(2.3).2

A. $\frac{16}{3}$

B. 1

C. 2

D. 4

E. 8

Question 240:

Solve $x^2 + 3x - 5 = 0$

A. $x = -\frac{3}{2} \pm \frac{\sqrt{11}}{2}$

B. $x = \frac{3}{2} \pm \frac{\sqrt{11}}{2}$

C. $x = -\frac{3}{2} \pm \frac{\sqrt{11}}{4}$

D. $x = \frac{3}{2} \pm \frac{\sqrt{11}}{4}$

E. $x = \frac{3}{2} \pm \frac{\sqrt{29}}{2}$

F. $x = -\frac{3}{2} \pm \frac{\sqrt{29}}{2}$

Question 241:

How many times do the curves $y = x^3$ and $y = x^2 + 4x + 14$ intersect?

A. 0

B. 1

C. 2

D. 3

E. 4

Question 242:

Which of the following graphs **do not** intersect?

1. $y = x$
2. $y = x^2$
3. $y = 1-x^2$
4. $y = 2$

A. 1 and 2

B. 2 and 3

C. 3 and 4

D. 1 and 3

E. 1 and 4

F. 2 and 4

Question 243:

Calculate the product of 897,653 and 0.009764.

A. 87646.8

B. 8764.68

C. 876.468

D. 87.6468

E. 8.76468

F. 0.876468

Question 244:

Solve for x: $\frac{7x + 3}{10} + \frac{3x + 1}{7} = 14$

A. $\frac{929}{51}$ B. $\frac{949}{47}$ C. $\frac{949}{79}$ D. $\frac{980}{79}$

Question 245:

What is the area of an equilateral triangle with side length x.

A. $\frac{x^2\sqrt{3}}{4}$ B. $\frac{x\sqrt{3}}{4}$ C. $\frac{x^2}{2}$ D. $\frac{x}{2}$ E. x^2 F. x

Question 246:

Simplify $3 - \frac{7x\left(25x^2 - 1\right)}{49x^2(5x+1)}$

A. $3 - \frac{5x-1}{7x}$

B. $3 - \frac{5x+1}{7x}$

C. $3 + \frac{5x-1}{7x}$

D. $3 + \frac{5x+1}{7x}$

E. $3 - \frac{5x^2}{49}$

F. $3 + \frac{5x^2}{49}$

Question 247:

Solve the equation $x^2 - 10x - 100 = 0$

A. $-5 \pm 5\sqrt{5}$

B. $-5 \pm \sqrt{5}$

C. $5 \pm 5\sqrt{5}$

D. $5 \pm \sqrt{5}$

E. $5 \pm 5\sqrt{125}$

F. $-5 \pm \sqrt{125}$

Question 248:

Rearrange $x^2 - 4x + 7 = y^3 + 2$ to make x the subject.

A. $x = 2 \pm \sqrt{y^3 + 1}$

B. $x = 2 \pm \sqrt{y^3 - 1}$

C. $x = -2 \pm \sqrt{y^3 - 1}$

D. $x = -2 \pm \sqrt{y^3 + 1}$

E. x cannot be made the subject for this equation.

Question 249:

Rearrange $3x + 2 = \sqrt{7x^2 + 2x + y}$ to make y the subject.

A. $y = 4x^2 + 8x + 2$

B. $y = 4x^2 + 8x + 4$

C. $y = 2x^2 + 10x + 2$

D. $y = 2x^2 + 10x + 4$

E. $y = x^2 + 10x + 2$

F. $y = x^2 + 10x + 4$

Question 250:

Rearrange $y^4 - 4y^3 + 6y^2 - 4y + 2 = x^5 + 7$ to make y the subject.

A. $y = 1 + (x^5 + 7)^{1/4}$

B. $y = -1 + (x^5 + 7)^{1/4}$

C. $y = 1 + (x^5 + 6)^{1/4}$

D. $y = -1 + (x^5 + 6)^{1/4}$

Question 251:

The aspect ratio of my television screen is 4:3 and the diagonal is 50 inches. What is the area of my television screen?

A. 1,200 inches^2

B. 1,000 inches^2

C. 120 inches^2

D. 100 inches^2

E. More information needed.

Question 252:

Rearrange the equation $\sqrt{1 + 3x^{-2}} = y^5 + 1$ to make x the subject.

A. $x = \frac{(y^{10} + 2y^5)}{3}$

B. $x = \frac{3}{(y^{10} + 2y^5)}$

C. $x = \sqrt{\frac{3}{y^{10} + 2y^5}}$

D. $x = \sqrt{\frac{y^{10} + 2y^5}{3}}$

E. $x = \sqrt{\frac{y^{10} + 2y^5 + 2}{3}}$

F. $x = \sqrt{\frac{3}{y^{10} + 2y^5 + 2}}$

Question 253:

Solve $3x - 5y = 10 \, and \, 2x + 2y = 13$.

A. $(x, y) = (\frac{19}{16}, \frac{85}{16})$

B. $(x, y) = (\frac{85}{16}, -\frac{19}{16})$

C. $(x, y) = (\frac{85}{16}, \frac{19}{16})$

D. $(x, y) = (-\frac{85}{16}, -\frac{19}{16})$

E. No solutions possible.

Question 254:

The two inequalities $x + y \leq 3 \; and \, x^3 - y^2 < 3$ define a region on a plane. Which of the following points is inside the region?

A. (2, 1)

B. (2.5, 1)

C. (1, 2)

D. (3, 5)

E. (1, 2.5)

F. None of the above.

Question 255:

How many times do $y = x + 4 \, and \, y = 4x^2 + 5x + 5$ intersect?

A. 0

B. 1

C. 2

D. 3

E. 4

Question 256:

How many times do $y = x^3 \, and \, y = x$ intersect?

A. 0

B. 1

C. 2

D. 3

E. 4

Question 257:

A cube has unit length sides. What is the length of a line joining a vertex to the midpoint of the opposite side?

A. $\sqrt{2}$

B. $\sqrt{\frac{3}{2}}$

C. $\sqrt{3}$

D. $\sqrt{5}$

E. $\frac{\sqrt{5}}{2}$

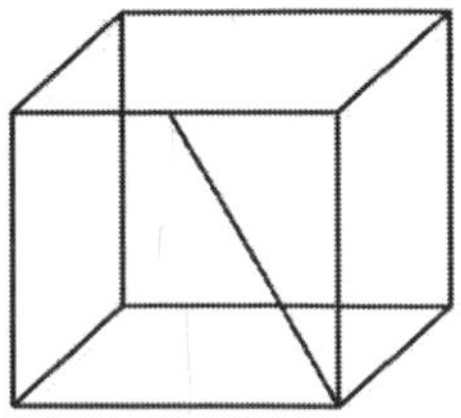

Question 258:

Solve for x, y, and z.

1. $x + y - z = -1$
2. $2x - 2y + 3z = 8$
3. $2x - y + 2z = 9$

	x	y	z
A	2	-15	-14
B	15	2	14
C	14	15	-2
D	-2	15	14
E	2	-15	14
F	No solutions possible		

Question 259:

Fully factorise: $3a^3 - 30a^2 + 75a$

A. $3a(a-3)^3$

B. $a(3a-5)^2$

C. $3a(a^2-10a+25)$

D. $3a(a-5)^2$

E. $3a(a+5)^2$

Question 260:

Solve for x and y:

$$4x + 3y = 48$$
$$3x + 2y = 34$$

	x	y
A	8	6
B	6	8
C	3	4
D	4	3
E	30	12
F	12	30
G	No solutions possible	

Question 261:

Evaluate: $\frac{-\left(5^2 - 4 \times 7\right)^2}{-6^2 + 2 \times 7}$

A. $-\frac{3}{50}$ B. $\frac{11}{22}$ C. $-\frac{3}{22}$ D. $\frac{9}{50}$ E. $\frac{9}{22}$ F. 0

Question 262:

All license plates are 6 characters long. The first 3 characters consist of letters and the next 3 characters of numbers. How many unique license plates are possible?

A. 676,000
B. 6,760,000
C. 67,600,000
D. 1,757,600
E. 17,576,000
F. 175,760,000

Question 263:

How many solutions are there for: $2(2(x^2 - 3x)) = -9$

A. 0
B. 1
C. 2
D. 3
E. Infinite solutions.

Question 264:

Evaluate: $\left(x^{\frac{1}{2}}\, y^{-3}\right)^{\frac{1}{2}}$

A. $\frac{x^{\frac{1}{2}}}{y}$
B. $\frac{x}{y^{\frac{3}{2}}}$
C. $\frac{x^{\frac{1}{4}}}{y^{\frac{3}{2}}}$
D. $\frac{y^{\frac{1}{4}}}{x^{\frac{3}{2}}}$

Question 265:

Bryan earned a total of £ 1,240 last week from renting out three flats. From this, he had to pay 10% of the rent from the 1-bedroom flat for repairs, 20% of the rent from the 2-bedroom flat for repairs, and 30% from the 3-bedroom flat for repairs. The 3-bedroom flat costs twice as much as the 1-bedroom flat. Given that the total repair bill was £ 276 calculate the rent for each apartment.

	1 Bedroom	2 Bedrooms	3 Bedrooms
A	280	400	560
B	140	200	280
C	420	600	840
D	250	300	500
E	500	600	1,000

Question 266:

Evaluate: $5\,[5\left(6^2 - 5 \times 3\right) + 400^{\frac{1}{2}}]^{1/3} + 7$

A. 0
B. 25
C. 32
D. 49
E. 56
F. 200

SECTION 1B: MATHS

Question 267:

What is the area of a regular hexagon with side length 1?

A. $3\sqrt{3}$

B. $\frac{3\sqrt{3}}{2}$

C. $\sqrt{3}$

D. $\frac{\sqrt{3}}{2}$

E. 6

F. More information needed

Question 268:

Dexter moves into a new rectangular room that is 19 metres longer than it is wide, and its total area is 780 square metres. What are the room's dimensions?

A. Width = 20 m; Length = -39 m

B. Width = 20 m; Length = 39 m

C. Width = 39 m; Length = 20 m

D. Width = -39 m; Length = 20 m

E. Width = -20 m; Length = 39 m

Question 269:

Tom uses 34 meters of fencing to enclose his rectangular lot. He measured the diagonals to 13 metres long. What is the length and width of the lot?

A. 3 m by 4 m

B. 5 m by 12 m

C. 6 m by 12 m

D. 8 m by 15 m

E. 9 m by 15 m

F. 10 m by 10 m

Question 270:

Solve $\frac{3x - 5}{2} + \frac{x + 5}{4} = x + 1$

A. 1

B. 1.5

C. 3

D. 3.5

E. 4.5

F. None of the above

Question 271:

Calculate: $\frac{5.226 \times 10^6 + 5.226 \times 10^5}{1.742 \times 10^{10}}$

A. 0.033

B. 0.0033

C. 0.00033

D. 0.000033

E. 0.0000033

Question 272:

Calculate the area of the triangle shown to the right:

A. $3 + \sqrt{2}$

B. $\frac{2 + 2\sqrt{2}}{2}$

C. $2 + 5\sqrt{2}$

D. $3 - \sqrt{2}$

E. 3

F. 6

$2 + \sqrt{2}$

$4 - \sqrt{2}$

Question 273:

Rearrange $\sqrt{\frac{4}{x} + 9} = y - 2$ to make x the subject.

A. $x = \frac{11}{(y-2)^2}$

B. $x = \frac{9}{(y-2)^2}$

C. $x = \frac{4}{(y+1)(y-5)}$

D. $x = \frac{4}{(y-1)(y+5)}$

E. $x = \frac{4}{(y+1)(y+5)}$

F. $x = \frac{4}{(y-1)(y-5)}$

Question 274:

When 5 is subtracted from 5x the result is half the sum of 2 and 6x. What is the value of x?

A. 0
B. 1
C. 2
D. 3
E. 4
F. 6

Question 275:

Estimate $\frac{54.98 + 2.25^2}{\sqrt{905}}$

A. 0
B. 1
C. 2
D. 3
E. 4
F. 5

Question 276:

At a Pizza Parlour, you can order single, double or triple cheese in the crust. You also have the option to include ham, olives, pepperoni, bell pepper, meat balls, tomato slices, and pineapples. How many different types of pizza are available at the Pizza Parlour?

A. 10
B. 96
C. 192
D. 384
E. 768
F. None of the above

Question 277:

Solve the simultaneous equations $x^2 + y^2 = 1$ and $x + y = \sqrt{2}$, for x, y > 0

A. $(x, y) = (\frac{\sqrt{2}}{2}, \frac{\sqrt{2}}{2})$

B. $(x, y) = (\frac{1}{2}, \frac{\sqrt{3}}{2})$

C. $(x, y) = (\sqrt{2} - 1, 1)$

D. $(x, y) = (\sqrt{2}, \tfrac{1}{2})$

Question 278:

Which of the following statements is **FALSE**?

A. Congruent objects always have the same dimensions and shape.

B. Congruent objects can be mirror images of each other.

C. Congruent objects do not always have the same angles.

D. Congruent objects can be rotations of each other.

E. Two triangles are congruent if they have two sides and one angle of the same magnitude.

Question 279:

Solve the inequality $x^2 \geq 6 - x$

A. $x \leq -3$ and $x \leq 2$

B. $x \leq -3$ and $x \geq 2$

C. $x \geq -3$ and $x \leq 2$

D. $x \geq -3$ and $x \geq 2$

E. $x \geq 2$ only

F. $x \geq -3$ only

Question 280:

The hypotenuse of an equilateral right-angled triangle is x cm. What is the area of the triangle in terms of x?

A. $\frac{x^2}{4}$ B. $\frac{x}{4}$ C. $\frac{3x^2}{4}$ D. $\frac{x^2}{10}$

Question 281:

Mr Heard derives a formula: Q= $\frac{(X+Y)^2 A}{3B}$. He doubles the values of X and Y, halves the value of A and triples the value of B. What happens to value of Q?

A. Decreases by $\frac{1}{3}$
B. Increases by $\frac{1}{3}$
C. Decreases by $\frac{2}{3}$
D. Increases by $\frac{2}{3}$
E. Increases by $\frac{4}{3}$
F. Decreases by $\frac{4}{3}$

Question 282:

Consider the graphs $y = x^2 - 2x + 3$, and $y = x^2 - 6x - 10$. Which of the following is true?

A. Both equations intersect the x-axis.
B. Neither equation intersects the x-axis.
C. The first equation does not intersect the x-axis; the second equation intersects the x-axis.
D. The first equation intersects the x-axis; the second equation does not intersect the x-axis.

Questions 283 -317 are more representative of the difficulty of questions you are likely to encounter.

Question 283:

The vertex of an equilateral triangle is covered by a circle whose radius is half the height of the triangle. What percentage of the triangle is covered by the circle?

A. 12%
B. 16%
C. 23%
D. 33%
E. 41%
F. 50%

Question 284:

Three equal circles fit into a quadrilateral as shown, what is the height of the quadrilateral?

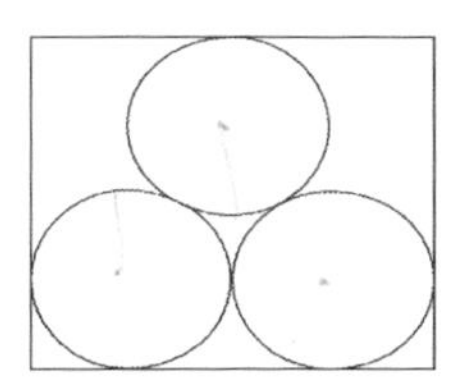

A. $2\sqrt{3}r$
B. $\left(2+\sqrt{3}\right)r$
C. $\left(4-\sqrt{3}\right)r$
D. $3r$
E. $4r$
F. More Information Needed

Question 285:
Two pyramids have equal volume and height, one with a square of side length $\boldsymbol{a}$ and one with a hexagonal base of side length $\boldsymbol{b}$. What is the ratio of the side length of the bases?

A. $\sqrt{\frac{3\sqrt{3}}{2}}$ B. $\sqrt{\frac{2\sqrt{3}}{3}}$ C. $\sqrt{\frac{3}{2}}$ D. $\frac{2\sqrt{3}}{3}$ E. $\frac{3\sqrt{3}}{2}$

Question 286:
One 9 cm cube is cut into 3 cm cubes. The total surface area increases by a factor of:

A. $\frac{1}{3}$ B. $\sqrt{3}$ C. 3 D. 9 E. 27

Question 287:
A cone has height twice its base width (four times the circle radius). What is the cone angle (half the angle at the vertex)?

A. 30°
B. $\sin^{-1}\left(\frac{r}{2}\right)$
C. $\sin^{-1}\left(\frac{1}{\sqrt{17}}\right)$
D. $\cos^{-1}(\sqrt{17})$

Question 288:
A hemispherical speedometer has a maximum speed of 200 mph. What is the angle travelled by the needle at a speed of 70 mph?

A. 28° B. 49° C. 63° D. 88° E. 92°

Question 289:
Two rhombuses, A and B, are similar. The area of A is 10 times that of B. What is the ratio of the smallest angles over the ratio of the shortest sides?

A. 0 B. $\frac{1}{10}$ C. $\frac{1}{\sqrt{10}}$ D. $\sqrt{10}$ E. ∞

Question 290:
$If f^{-1}(-x) = \ln(2x^2)$ what is $f(x)$?

A. $\sqrt{\frac{e^y}{2}}$ B. $\sqrt{\frac{e^{-y}}{2}}$ C. $\frac{e^y}{2}$ D. $\frac{-e^y}{2}$ E. $\sqrt{\frac{e^y}{2}}$

Question 291:
Which of the following is largest for $0 < x < 1$

A. $log_8(x)$ B. $log_{10}(x)$ C. e^x D. x^2 E. $\sin(x)$

Question 292:
x is proportional to y cubed, y is proportional to the square root of z. $x \propto y^3,\ y \propto \sqrt{z}$.
If z doubles, x changes by a factor of:

A. $\sqrt{2}$ B. 2 C. $2\sqrt{2}$ D. $\sqrt[3]{4}$ E. 4

Question 293:
The area between two concentric circles (shaded) is three times that of the inner circle.

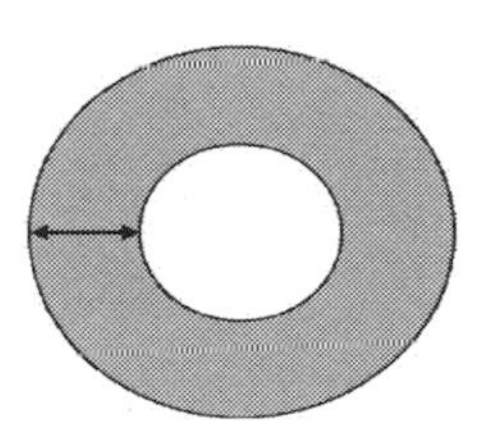

What's the size of the gap?

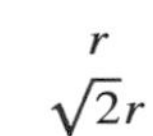

A. r B. $\sqrt{2}r$ C. $\sqrt{3}r$ D. $2r$ E. $3r$ F. $4r$

Question 294:
Solve $-x^2 \leq 3x - 4$

A. $x \geq \frac{4}{3}$

B. $1 < x < 4$

C. $x \leq 2$

D. $x \geq 1$ or $x \geq -4$

E. $-1 \leq x \leq \frac{3}{4}$

Question 295:
The volume of a sphere is numerically equal to its projected area. What is its radius?

A. $\frac{1}{2}$ B. $\frac{2}{3}$ C. $\frac{3}{4}$ D. $\frac{4}{3}$ E. $\frac{3}{2}$

Question 296:
What is the range where $x^2 < \frac{1}{x}$?

A. $x < 0$

B. $0 < x < 1$

C. $x > 0$

D. $x \geq 1$

E. *None*

Question 297:
Simplify and solve: (e - a) (e + b) (e – c) (e + d)…(e - z)?

A. 0

B. e^{26}

C. e^{26} (a-b+c-d...+z)

D. e^{26} (a+b-c+d...-z)

E. e^{26} (abcd...z)

F. None of the above.

Question 298:
Find the value of k such that the vectors $a = -i + 6j$ and $b = 2i + kj$ are perpendicular.

A. -2 B. $-\frac{1}{3}$ C. $\frac{1}{3}$ D. 2

Question 299:
What is the perpendicular distance between point $\boldsymbol{p}$ with position vector $4\boldsymbol{i} + 5\boldsymbol{j}$ and the line L given by vector equation $\boldsymbol{r} = -3\boldsymbol{i} + \boldsymbol{j} + \lambda(\boldsymbol{i} + 2\boldsymbol{j})$

A. $2\sqrt{7}$ B. $5\sqrt{2}$ C. $2\sqrt{5}$ D. $7\sqrt{2}$

Question 300:

Find k such that point $\begin{pmatrix} 2 \\ k \\ -7 \end{pmatrix}$ lies within the plane $\boldsymbol{r} = \begin{pmatrix} 2 \\ 3 \\ -1 \end{pmatrix} + \lambda \begin{pmatrix} 4 \\ 1 \\ 0 \end{pmatrix} + \mu \begin{pmatrix} 2 \\ 1 \\ 3 \end{pmatrix}$

A. -2
B. -1
C. 0
D. 1
E. 2

Question 301:
What is the largest solution to $\sin(-2\theta) = 0.5$ for $\frac{\pi}{2} \leq x \leq 2\pi$?

A. $\frac{5\pi}{3}$ B. $\frac{4\pi}{3}$ C. $\frac{5\pi}{6}$ D. $\frac{7\pi}{6}$ E. $\frac{11\pi}{6}$

Question 302:
$\cos^4(x) - \sin^4(x) \equiv$

A. $cos(2x)$
B. $2cos(x)$
C. $sin(2x)$
D. $sin(x)cos(x)$
E. $tan(x)$

Question 303:
How many real roots does $y = 2x^5 - 3x^4 + x^3 - 4x^2 - 6x + 4$ have?

A. 1
B. 2
C. 3
D. 4
E. 5

Question 304:

What is the sum of 8 terms, $\sum_{1}^{8} u_n$, of an arithmetic progression with $u_1 = 2$ and $d = 3$.

A. 15
B. 82
C. 100
D. 184
E. 282

Question 305:
What is the coefficient of the x^2 term in the binomial expansion of $(2 - x)^5$?

A. -80 B. 48 C. 40 D. 48 E. 80

Question 306:
Given you have already thrown a 6, what is the probability of throwing three consecutive 6s using a fair die?

A. $\frac{1}{216}$ B. $\frac{1}{36}$ C. $\frac{1}{6}$ D. $\frac{1}{2}$ E. 1

Question 307:

Three people, A, B and C play darts. The probability that they hit a bullseye are respectively $\frac{1}{5}, \frac{1}{4}, \frac{1}{3}$. What is the probability that at least two shots hit the bullseye?

A. $\frac{1}{60}$ B. $\frac{1}{30}$ C. $\frac{1}{12}$ D. $\frac{1}{6}$ E. $\frac{3}{20}$

Question 308:
If probability of having blonde hair is 1 in 4, the probability of having brown eyes is 1 in 2 and the probability of having both is 1 in 8, what is the probability of having neither blonde hair nor brown eyes?

A. $\frac{1}{2}$ B. $\frac{3}{4}$ C. $\frac{3}{8}$ D. $\frac{5}{8}$ E. $\frac{7}{8}$

Question 309:
Differentiate and simplify $y = x(x+3)^4$

A. $(x+3)^3$
B. $(x+3)^4$
C. $x(x+3)^3$
D. $(5x+3)(x+3)^3$
E. $5x^3(x+3)$

Question 310:

Evaluate $\int_1^2 \frac{2}{x^2} dx$

A. -1 B. $\frac{1}{3}$ C. 1 D. $\frac{21}{4}$ E. 2

Question 311:
Express $\frac{5i}{1+2i}$ in the form $a + bi$

A. $1 + 2i$
B. $4i$
C. $1 - 2i$
D. $2 + i$
E. $5 - i$

Question 312:

Simplify $7\log_a(2) - 3\log_a(12) + 5\log_a(3)$

A. $log_{2a}(18)$

B. $log_a(18)$

C. $log_a(7)$

D. $9log_a(17)$

E. $-log_a(7)$

Question 313:

What is the equation of the asymptote of the function $y = \dfrac{2x^2 - x + 3}{x^2 + x - 2}$

A. $x = 0$

B. $x = 2$

C. $y = 0.5$

D. $y = 0$

E. $y = 2$

Question 314:

Find the intersection(s) of the functions $y = e^x - 3$ and $y = 1 - 3e^{-x}$

A. 0 and ln(3)

B. 1

C. In(4) and 1

D. In(3)

Question 315:

Find the radius of the circle $x^2 + y^2 - 6x + 8y - 12 = 0$

A. 3

B. $\sqrt{13}$

C. 5

D. $\sqrt{37}$

E. 12

Question 316:

What value of ***a*** minimises $\int_0^a 2\sin(-x)dx$?

A. 0.5π

B. π

C. 2π

D. 3π

E. 4

Question 317:

When $\dfrac{2x + 3}{(x - 2)(x - 3)^2}$ is expressed as partial fractions, what is the numerator in the $\dfrac{A}{(x - 2)}$ term:

A. -7

B. -1

C. 3

D. 6

E. 7

END OF SECTION

SECTION 1B: ADVANCED MATHS

SECTION 2: Writing Task

The Basics

In section 2, you have to write an essay based upon a passage. **There is no choice of essay title** meaning that you have to do the question that comes up. Whilst different questions will inevitably demand differing levels of comprehension and knowledge, it is important to realise that one of the major skills being tested is actually your ability to construct a logical and coherent argument- and to convey it to the lay-reader.

Section 2 of the ECAA is frequently neglected by lots of students, who choose to spend their time on section 1 instead. However, it is possible to rapidly improve in it and given that it may come up at your interview, well worth the time investment!

The aim of section 2 is not to write as much as you can. Rather, the examiner is looking for you to make interesting and well supported points, and tie everything neatly together for a strong conclusion. Make sure you're writing critically and concisely; not rambling on. **Irrelevant material can actually lower your score.**

Essay Structure

Basic Structure

ECAA Essays should follow the standard format of Introduction [?] Main Body [?] Conclusion.

The introduction should be the smallest portion of the essay (no more than one small paragraph) and be used to provide a smooth segue into the rather more demanding "argue for/against" part of the question. This main body requires a firm grasp of the concept being discussed and the ability to strengthen and support the argument with a wide variety of examples from multiple fields. This section should give a balanced approach to the question, exploring **at least two distinct ideas**. Supporting evidence should be provided throughout the essay, with examples referred to when possible.

The concluding final part effectively is a chance for you to shine- be brave and make an **innovative yet firmly grounded conclusion** for an exquisite mark. The conclusion should bring together all sides of the argument, in order to reach a clear and concise answer to the question. There should be an obvious logical structure to the essay, which reflects careful planning and preparation.

Paragraphs

Paragraphs are an important formatting tool which show that you have thought through your arguments and are able to structure your ideas clearly. A new paragraph should be used every time a new idea is introduced. There is no single correct way to arrange paragraphs, but it's important that each paragraph flows smoothly from the last. A slick, interconnected essay shows that you have the ability to communicate and organise your ideas effectively.

Remember- the emphasis should remain on quality and not quantity. An essay with fewer paragraphs, but with well-developed ideas, is much more effective than a number of short, unsubstantial paragraphs that fail to fully grasp the question at hand.

Planning

Why should I plan my essay?

The vast **majority of problems are caused by a lack of planning** - usually because students just want to get writing as they are worried about finishing on time. Fourty minutes is long enough to be able to plan your essay well and *still* have time to write it so don't feel pressured to immediately start writing.

There are multiple reasons you should plan your essay for the first 5-10 minutes of section 2:

➢ It allows you to get all your thoughts ready before you put pen to paper.

➢ You'll write faster once you have a plan.

➢ You run the risk of missing the point of the essay or only answering part of it if you don't plan adequately.

How much time should I plan for?

There is no set period of time that should be dedicated to planning, and everyone will dedicate a different length of time to the planning process. You should spend as long planning your essay as you require, but it is essential that you leave enough time to write the essay. As a rough guide, it is **worth spending about 5-10 minutes to plan** and the remaining time on writing the essay. However, this is not a strict rule, and you are advised to tailor your time management to suit your individual style.

How should I go about the planning process?

There are a variety of methods that can be employed in order to plan essays (e.g. bullet-points, mind-maps etc). If you don't already know what works best, it's a good idea to experiment with different methods.

Generally, the first step is to gather ideas relevant to the question, which will form the basic arguments around which the essay is to be built. You can then begin to structure your essay, including the way that points will be linked. At this stage it is worth considering the balance of your argument, and confirming that you have considered arguments from both sides of the debate. Once this general structure has been established, it is useful to consider any examples or real world information that may help to support your arguments. Finally, you can begin to assess the plan as a whole, and establish what your conclusion will be based on your arguments.

Introduction

Why are introductions important?
An introduction provides tutors with their first opportunity to examine your work. The introduction is where first impressions are formed, and these can be extremely important in producing a convincing argument. A well-constructed introduction shows that you have really thought about the question, and can indicate the logical flow of arguments that is to come.

What should an introduction do?
A good introduction should **briefly explain the statement or quote** and give any relevant background information in a concise manner. However, don't fall into the trap of just repeating the statement in a different way. The introduction is the first opportunity to suggest an answer to the question posed- the main body is effectively your justification for this answer.

Main Body

How do I go about making a convincing point?
Each idea that you propose should be supported and justified, in order to build a convincing overall argument. A point can be solidified through a basic Point [?] Evidence [?] Evaluation process. By following this process, you can be assured each sentence within a paragraph builds upon the last, and that all the ideas presented are well solidified.

How do I achieve a logical flow between ideas?
One of the most effective ways of displaying a good understanding of the question is to keep a logical flow throughout your essay. This means linking points effectively between paragraphs, and creating a congruent train of thought for the examiner as the argument develops. A good way to generate this flow of ideas is to provide ongoing comparisons of arguments, and discussing whether points support or dispute one another.

Should I use examples?
In short – yes! Examples can help boost the validity of arguments, and can help display high quality writing skills. Examples can add a lot of weight to your argument and make an essay much more relevant to the reader. When using examples, you should ensure that they are relevant to the point being made, as they will not help to support an argument if they are not.

Some questions will provide more opportunities to include examples than others so don't worry if you aren't able to use as many examples as you would have liked. There is no set rule about how many examples should be included!

Top tip! Remember that there is no single correct answer to these questions and you're not expected to be able to fit everything onto one page. Instead it's better to pick a few key points to focus on.

Conclusion

The conclusion provides an opportunity to emphasise the **overall sentiment of your essay** which readers can then take away. It should summarise what has been discussed during the main body and give a definitive answer to the question.

Some students use the conclusion to **introduce a new idea that hasn't been discussed**. This can be an interesting addition to an essay, and can help make you stand out. However, it is by no means, a necessity. In fact, a well-organised, 'standard' conclusion is likely to be more effective than an adventurous but poorly executed one.

Common Mistakes

Ignoring the other side of the argument

You need to ensure that you show an appreciation for the fact that there are often two sides to the argument. Where appropriate, you should outline both points of view and how they pertain to the essay's main principles and then come to a reasoned judgement.

A good way to do this is to propose an argument that might be used against you, and then to argue why it doesn't hold true or seem relevant. You may use the format: *"some may say that...but this doesn't seem to be important because..."* in order to dispel opposition arguments, whilst still displaying that you have considered them. For example, *"some may say that fox hunting shouldn't be banned because it is a tradition. However, witch hunting was also once a tradition – we must move on with the times"*.

Missing Topic Sentences

A reader who is pressed for time should be able to read your introduction, the first line of every paragraph and your conclusion and be able to follow your argument. The filling of a paragraph will elaborate your point with examples. But the first sentence of the paragraph should provide the headline point.

- *Use topic sentences as punchy summaries for the theme of each paragraph*
- *Include a clear summary of the structure of your essay in your introduction*
- *Summarize briefly the theme of your points in your conclusion*
- *Ensure your conclusion also tells the reader your final decision*

Undefined Terms

Debates can be won or lost on the basis of the interpretation of a key term; ensure your interpretation of the key words is clearly explained. For example: "Does science or art shape our world?" Here, your interpretation of what it means to *shape* something is absolutely crucial to lay out before you start writing, so that your reader knows the scope of your argument. If *shape* to you means invent something new (like a potter shaping a pot out of a lump of clay), state this. But if you interpret *shape* to mean a gentle guide or influence on something, state that. You can then be more focused and precise in your discussion. Likewise, for this title ensure you are clear about the scope of what is science and what is art.

- *Define the key terms within the particular context of the question*
- *Be clear about your understanding of the scope*

No Sign-Posting

There is a delight to enjoying a long journey if you know (1) where you are going, (2) what you will see on the way and (3) how long it will take to get there. For the reader of your essay, the same logic applies. State briefly but clearly in the final sentence of your introduction the topics you will cover (preferably in the order you will cover them!). You don't need to give the entire game away (don't necessarily tell your reader precisely what your 'wow-factor' will be) but you can give them a solid hint as to your final destination. For example, "Having discussed these arguments in favour and against fox hunting, we conclude with a consideration of the wider issue of the role of governmental institutions in condoning and condemning the traditional pursuits of citizens." It is sometimes tempting to try to surprise your reader with an unexpected twist but this is not best practice for an academic essay.

- *Don't surprise your reader with unexpected twists in the main essay*
- *Do be clear in your introduction about the number of points you will make*
- *Do include your points in the order they will appear*

Long Introductions

Some students can start rambling and make introductions too long and unfocussed. Although background information about the topic can be useful, it is normally not necessary. Instead, the **emphasis should be placed on responding to the question**. Some students also just **rephrase the question** rather than actually explaining it. The examiner knows what the question is, and repeating it in the introduction is simply a waste of space in an essay where you are limited to just one A4 side.

Not including a Conclusion

An essay that lacks a conclusion is incomplete and can signal that the answer has not been considered carefully or that your organisation skills are lacking. **The conclusion should be a distinct paragraph** in its own right and not just a couple of rushed lines at the end of the essay.

Sitting on the Fence

Students sometimes don't reach a clear conclusion. You need to **ensure that you give a decisive answer to the question** and clearly explain how you've reached this judgement. Essays that do not come to a clear conclusion generally have a smaller impact and score lower.

Conclusions with no 'Wow-Factor'

Try to 'zoom out' in your conclusion, rather than merely summarising the points you have made and deciding that one set outweighs the other. Put the question back in a wider context, so that your decision has a wow-factor for why it really matters. For instance, if you have answered the question, "Is world peace achievable?" and you think it isn't, say why this matters. For example: "In an age of nuclear capability, attempts to achieve the impossible is a waste of scarce resources, so we'd be better off focusing policy and diplomacy on building safety nets to prevent escalations of inevitable conflicts into another world war."

- *Don't only repeat your arguments again in your conclusion*
- *Don't sit on the fence in your conclusion*
- *Do use the conclusion to zoom out for the final punchline: why does this matter?*

Missing the Point

Ensure you have identified what you think the 'Turning Point' of the question is, before you start writing. Within the title, which may be long and literary, identify the single core issue for you that you will discuss. For example, with the question, "Has the "digital age" destroyed the human right to anonymity?", restate it as a simple statement: the key question is whether previous to the introduction of digital technology we had a human right to anonymity which has now disappeared. You can then anchor your argument clearly on whether such a right had always existed before (perhaps so, perhaps not) and whether it has now disappeared (if it ever existed). By restating the key question, you will auto-generate a clear structure for yourself to follow.

- *Work out the hinge of the question before you start writing and state it clearly*

Worked Essay Questions

Passage 1

Despite the fact that some associate musicals with cheesy joy, the genre is not limited to gleeful stories, as can be demonstrated by the macabre musical, 'Sweeney Todd'. The original story of the murderous barber appears in a Victorian penny dreadful, 'The String of Pearls: A Romance'. The penny dreadful material was adapted for the 19th century stage, and in the 20th century was adapted into two separate melodramas, before the story was taken up by Stephen Sondheim and Hugh Wheeler. The pair turned it into a new musical, which has since been performed across the globe and been adapted into a film starring Johnny Depp.

Sondheim and Wheeler's drama tells a disturbing narrative: the protagonist, falsely accused of a crime by a crooked judge, escapes from Australia to be told that his wife was raped by that same man of the court. In response, she has committed suicide, and her daughter - Todd's daughter - has been made the ward of the judge. The eponymous figure ultimately goes on a killing spree, vowing vengeance against the people who have wronged him but also declaring 'we all deserve to die', and acting on this belief by killing many of his clients; men who come to his barbershop. His new partner in crime, Mrs Lovett, comes up with the idea of turning the bodies of his victims into the filling of pies, as a way of sourcing affordable meat - after all, she claims, 'times is hard'.

Cannibalism, vengeance, murder, and corruption - these are all themes that demonstrate that this show does not conform to a happy-clappy preconception of its genre.

Sondheim and Wheeler's musical has been adapted into a number of formats over the years, including the film 'Sweeney Todd: The Demon Barber of Fleet Street' directed by Tim Burton. The nature of a film production necessitated a number of changes to the musical. Burton even acknowledged that while it was based on the musical, they were out to make a film and not a Broadway show. Accordingly, a three-hour musical was cut into a two-hour film, which brought a number of challenges: some of the songs and the romance between Todd's daughter and Anthony (a sailor) had to be removed.

There was initially concern though as the film actors, while critically acclaimed in their profession, were not professional singers. However, that turned out to be a non-issue as the film's soundtrack received glowing reviews, in particular, Depp's voice which received positive critical appraisals.

QUESTION
"Musicals inappropriately make light of serious themes."
Discuss with reference to the passage above.

Example Plan

Introduction:

- Define "inappropriate". This could mean improper or obscene, but we take it here to mean a treatment that is incorrect given the content of the topic.
- The key question: does the rendering of a serious theme into musical theatre make it too light-hearted?

Paragraph 1:

- The genre of musical theatre is attended by people who expect feel-good entertainment. Thus themes of cannibalism, vengeance, murder, and corruption become part of the jokey, jovial entertainment atmosphere.
- *Passage Example:* The passage says musicals are associated with "cheesy joy" and "happy-clappy preconceptions." The true story of injustice and vengeance is thus trivialised by such a setting and the expectations of the audience in how they are supposed to react.

Paragraph 2:

- Bringing a story to life through musical theatre makes it memorable, and not necessarily frivolous. Songs are unique in their ability to highlight a theme clearly through not only words, but melodies too.
- *Passage Example*: "The original story of the murderous barber appears in a Victorian penny dreadful." The original story was a Victorian entertainment series, with the gory themes shared with readers in bite sized pieces, like a song telling the next stage in a musical.
- *Other Examples*: Victor Hugo's *Les Miserables* is a catchy and entertaining musical theatre piece, but would not be considered frivolous.

Paragraph 3:

- The genre could highlight the darkness of the themes through its juxtaposition against the musical theatre tradition. The friction between the jovial singing and the commentary on man's quest for revenge is startling and therefore stronger than when just expressed in words.
- *Passage Example*: The songs about "we all deserve to die" and "times is hard" are sung as a reasonable message within a horrifying context: a barber murdering clients and a pie-maker making human meat pies. The discomfort of the audience seeing the desperation of the characters and their evil reactions to it is increased by the fact that these are sung in catchy tunes.

Conclusion

- Summarize the points:
    - The feel-good entertainment genre does not offer a space for serious reflection.
    - Putting stories to songs makes the story, and therefore its content, memorable.
    - The choice of the genre increases the discomfort of the audience, rather than removes it.
- Now decide for yourself one way or another…
- Zoom out and say why the question really matters (just pick one to give your essay the 'wow-factor'):
    - Genre matters to complement the message of a story (if the story tells the tale of two characters facing difficulties and reacting in a horrific manner, the genre can also play a role in increasing the shock or discomfort).
    - Genre must match the authors' intentions (first published as a Penny Dreadful, the 'stage' version of this is naturally suited to musical adaptation).

Passage 2

Gutenberg's father was a man of good family. Very likely the boy was taught to read. But the books from which he learned were not like ours; they were written by hand. A better name for them than books is 'manuscripts,' which means handwritings.

While Gutenberg was growing up, a new way of making books came into use, which was a great deal better than copying by hand. It was what is called block printing. The printer first cut a block of hardwood the size of the page that he was going to print. Then he cut out every word of the written page upon the smooth face of his block. This had to be very carefully done. When it was finished, the printer had to cut away the wood from the sides of every letter. This left the letters raised, as the letters are in books now printed for the blind. The block was now ready to be used. The letters were inked, the paper was laid upon them and pressed down. With blocks, the printer could make copies of a book a great deal faster than a man could write them by hand. But the making of the blocks took a long time, and each block would print only one page.

Gutenberg enjoyed reading the manuscripts and block books that his parents and their wealthy friends had, and he often said it was a pity that only rich people could own books. Finally, he determined to contrive an easy and quick way of printing.

Gutenberg, indeed, found this way and made the first movable-type printing press in Europe, with pieces from lead, tin, and antimony. Crucially. Gutenberg's innovation was confirmed by the production of the 'Gutenberg Bible'; the first major book printed with the movable-type printing press. With the invention came cheaper and higher quality books, thereby encouraging the development of printing presses across Europe, which in turn increased the number of books in supply. A new feature of the movable-type printing press was the advent of oil-based ink, which was more durable. This has been termed the 'Gutenberg Revolution'.

QUESTION

"The maker and the author are equally important creators of a book."
Discuss with reference to the passage above.

Example Plan

Introduction:

- Define "Maker". In the case of this passage, we're referring to the manufacturer (the manuscript copier, the block printer and the printing press). "Creator": someone who makes something new.
- The key question: The maker and the author are important to whom? To the *reader*. What is important to the reader of a book, and how do the maker and author each respond to the reader's needs?

Paragraph 1:

- The result of more efficient technology is wider readership, and a book needs to reach the hands of a reader to fulfil its purpose (being read!). Without the maker, the book cannot be created and cannot be enjoyed.
- *Passage Example:* The passage says, "with the invention came cheaper and higher quality books ... which in turn increased the number of books in supply."
- *Passage Example*: "He often said it was a pity that only rich people could own books"; this tells us, prior to the makers of the books becoming more efficient, the books' readerships were sparse.

Paragraph 2:

- Copying cannot be considered creating. The content is what makes a book good for a reader; the way the book was made just uses the most efficient book-making technology of the time. The mechanisation of the book making process does not in any way alter the story within the book.
- *Passage Example*: We move from "copying by hand" to the printing press which is focused on economic optimisation, making "cheaper and higher quality books" eventually with "oil-based ink, which was more durable." This assists the function, but does not alter the content of the book and therefore the maker is less important to the reader than the author.

Paragraph 3:

- The way a book is made is part of how we experience it. Even in the modern day, when most books are printed in a similar fashion, the cover of a book is important to its reception: the choice of reviews on the back, the font size, choice of pictures...
- *Passage Example*: Gutenberg's books as a child "were written by hand": the handwriting of the copyist would be centre stage to the reader of a manuscript.
- *Passage Example:* We refer to the "Gutenberg Bible" because it was made by Gutenberg's technology. This shows that we highly acknowledge the maker of a book because it shapes our experience of reading it.

Paragraph 4:

- Changing the method of making a book changes its readership, which empowers people with more information with which to shape their world
- *Passage Example*: Gutenberg's book making process is called the "Gutenberg Revolution". The word revolution suggests a change in social systems: this is a creative process undertaken by the books' readers.

Conclusion

- Summarize the points:
    - Without a maker, a book is not read and therefore cannot fulfil its creative mission.
    - The technology – process of making a book – does not alter the story's content.
    - The technology does alter the way we read it because it changes our experience of reading.
    - The maker of a book can equip more readers to shape their world.
- Zoom out and say why the question really matters (just pick one to give your essay the 'wow-factor'):
    - Revolution: Writings have played centre stage in social critiques, commentaries and calls-to-arms. Revolution changes and creates societies, and the makers of books play a role in this.
    - Creation: The act of creation requires invention, which does not rely on readers or makers but instead its originator, the author.
    - Time: old books or books with previous owners' scribbling alter the reader's experience. Through time, people handling and interacting with a book alters its reception. The maker is the first in a line of people to start this dialogue with the current reader.

Passage 3 – The English Reformation

In the early 1500s, King Henry VIII set the English Church on a different course forever. Henry was undoubtedly a devout catholic when he took the throne. Indeed, he was a staunch defender of Catholicism in the face of threats from religious reformers, such as Luther. Impressed by Henry VIII's defence, the Pope gave him the title 'Defender of the Faith'. So how did Henry come to separate from the Roman Church?

Although historians are not universally in agreement, many put Henry VIII as the key driver behind separating the Church of England from Rome. Henry was disappointed in his marriage with Catherine of Aragon as, in spite of multiple pregnancies, they only had one daughter together. Henry though was desperate to conceive a son. He had a monumental ego and was, thus, concerned about his legacy. In order to secure his dynasty and ensure that the Tudor reign remained strong, he needed a legitimate son. Accordingly, he was eager to secure a divorce with his current wife and marry Anne Boleyn with the aim of having a legitimate son with her. The English church was under the authority of the Roman Catholic Church (of whom the Pope was the leader) and in order to separate from Catherine, Henry needed to obtain an annulment from the Pope. Despite the mammoth efforts of Henry's right-hand man, he was unable to secure an annulment of the marriage from Rome, which would have been the straightforward option. It became clear that Rome was not going to budge on this and from then, Henry began to pursue a separation from the Roman Church.

Historians also point to another reason for Henry's desire to break away from Rome. He liked the idea of being the only head of the church and the supreme leader. His ego influenced many of his key decisions, such as engaging in wars abroad, and this decision was no different.

A number of historians suggest that Thomas Cromwell was the man behind the separation. Indeed, Cromwell played a significant role in engineering it. With control of the King's parliamentary affairs, he persuaded Parliament to enact a supplication pronouncing Henry as 'the only head' of the church, establishing the doctrine of royal supremacy. This was in clear conflict with Papal authority and began the process of breaking away from the Roman Church. But while it is clear that Cromwell had a vital role in the break from Rome, the obvious must still be repeated – were it not for Henry's desire of a break, there would not have been such a break.

Through a series of Acts of Parliament over two years, the break from Rome was secured and ties between the English church and Rome were severed. One such Act of Parliament in 1934, the Act of Supremacy, declared the King as 'the only Supreme head in earth of the Church of England.' This drastic change put the English church on a new course and while there were no major day-to-day changes initially, it planted the seed for the differences we see today between the Roman Catholic Church and the Church of England.

QUESTION
"A King's ego is the primary driver of change in a monarchy."
Discuss with reference to the passage above.

Example Plan

Introduction:

- Define monarchy, ego and what we mean by 'change': change to what? To legislation, to international relations, to society's norms.
- The key question is whether there are stronger drivers of change within a country: we must identify different sources of pressure for change, and evaluate their relative strength to a king's pressure. We look at this in the specific context of Tudor England.

Paragraph 1:

- Despite the authority of the Roman Catholic Church in national matters relating not only to religious practice, Henry VIII was able to change the religion of his entire nation to suit his aims.
- *Passage Example:* "Many put Henry VIII as the key driver behind separating the Church of England from Rome." His ambitions for his royal line to continue are what motivate the Reformation.

Paragraph 2:

- The will of a King is a necessary but not sufficient condition for change. Henry VIII required intelligent advisors to enact his changes; without these his ego alone would not have triumphed.
- *Passage Example*: "Thomas Cromwell was the man behind the separation". A wish for something from a King is insufficient to being that change about.
- *Passage Example:* "Cromwell had a vital role in the break from Rome."

Paragraph 3:

- Henry VIII's advisors were responding to his commands, not acting as independent advisors; when they did not achieve his will, he found another way.
- *Passage Example:* Henry VIII's advisor made "mammoth efforts … to secure an annulment of the marriage" … and when he failed, Henry pursued another course of action.
- *Passage Example:* Thomas Cromwell was acting on Henry's wishes: "were it not for Henry's desire for a break, there would not have been such a break."

Paragraph 4:

- The representatives from Rome, the King's advisors and the Parliament are all complicit in not exerting sufficient levels of their own pressure to counteract the pressure from the King's ego. This may not be the case always: the King's wishes need official stamps of approval from many sources for a change to be enacted.
- *Passage Example*: "Through a series of Acts of Parliament over two years…" This comes after the debates with Rome and the research of the advisors. It was these acts that solidified the King's wishes.

Conclusion

- Summarize the points:
    - Henry VIII was successful, despite the Catholic Church's efforts, to make himself "Supreme head in earth of the Church of England". This is a triumph of the King's ego.
    - The King's ego needed advisors to carry out his wishes; without this, he would have had no choice but to bow to Rome.
    - When an ego is strong enough, the advisors find a way to please it or they are replaced.
    - The actual change came from laws made in Parliament; should these have been rejected, Henry VIII's ego would have not brought about change, only conflict.
- Now decide for yourself one way or another…
- Zoom out and say why the question really matters (just pick one to give your essay the 'wow-factor'):
    - The King is only a figurehead for a set of ruling institutions; an individual cannot enact change on their own.
    - The King initiates a drive for change. The relative power of his advisors and his constraining institutions determine whether the change will be put in effect, but his will is the catalyst.

Passage 4

Most institutions in the country are businesses – shops, factories, energy companies, airlines, and train companies, to name a few types. They are the bedrock of society, employ most people in it and it is, thus, crucial that we examine their values.

The overriding objective of businesses is to make the most profit (i.e. maximise on revenue and minimise on costs). The notion was first popularly expounded by Adam Smith in his book, 'The Wealth of Nations' in 1776. Furthermore, his view was that if an individual considers merely their own interests to create and sell goods or services for the most profit, the invisible hand of the market will lead that activity to maximise the welfare of society. For example, in order to maximise profits, sellers will only produce and sell goods that society wants. If they try to sell things people don't want, no one would buy it. This is how the free market works. Indeed, the focus on profit is the basis on which companies operate and encourages them to innovate and produce goods that consumers want, such as iPhones and computers. So there are clear benefits to the profit maximisation theory.

This is a more effective society than, for example, a communist society where the government decides what to produce – as the government has no accurate way of deciding what consumers need and want. Arguably, the poverty that communist regimes such as the Soviet Union created have instilled this notion further.

However, were companies left to their own devices to engage in profit maximisation, what would stop them from exploiting workers? What would stop them from dumping toxic chemicals into public rivers? Engaging in such practices would reduce their costs of production, which would increase their profits. However, this would be very damaging to the environment. Accordingly, other objectives should be relevant. Businesses can also do other bad things to make a profit as well. For example, selling products to people who don't want or need them.

Corporate social responsibility entails other possible objectives for businesses, such as a consideration of the interests of stakeholders. A stakeholder is, in essence, anyone who is significantly affected by a company decision, such as employees or the local community. One business decision can have huge impacts on stakeholders. For example, a decision to transfer a call centre from the UK to India would likely increase profits, as wage costs for Indian workers can be much lower than that of British workers. This increase in profits would benefit the shareholders, however, it negatively harms other stakeholders. It would make many employees redundant. Here, there is arguably a direct conflict between profit maximisation and employees' interest. Nonetheless, moving call centres abroad does not always work. Given the different cultures and accents, companies have received complaints from frustrated customers. This, in fact, led BT to bring back a number of call centres to the UK.

However, the objective of profit maximisation has not always led to maximum welfare for society. Arguably, as banks sought to maximise their profits, they lent money to individuals who could not afford to pay it back. Eventually, many borrowers stopped meeting their repayments and lost banks enormous amounts. This led to a need for banks to be bailed out by the government and Lehman Brothers; one of the largest US banks that collapsed. Arguably, though, this was more due to idiocy rather than profit maximisation alone – in the end, the banks lost billions.

QUESTION

"Enforcing Corporate Social Responsibility interferes with the market's ability to give people what they want." Discuss with reference to the passage above.

Example Plan

Introduction:

- Define Corporate Social Responsibility (from the passage). The market: here, we mean the goods and services offered for sale by businesses and bought by consumers. People: here we mean both the consumers in the market and those not participating in the market.
- The key question: If businesses are driven by the objective to maximise profits, does the market truly give people what they want?

Paragraph 1:

- Adam Smith's invisible hand tells us that the market will allocate resources to maximise the welfare of society. If goods were produced and not demanded, this would waste resources. This cannot happen as the resources used to make these goods would be allocated to make other, demanded, goods.
- *Passage Example:* "Sellers will only produce and sell goods that society wants." The businesses may be maximising profits, but they can only do so within the constraints of selling items that are wanted by the people. Thus, what is on offer in a market is determined by the people's wants.

Paragraph 2:

- We need Corporate Social Responsibility to prevent the profit maximisation of businesses affecting stakeholders who are not necessarily the direct consumer of the good or service being produced. Their loss may outweigh the gains of those engaged in the transaction and so we have a net loss,
- *Passage Example*: "Corporate social responsibility entails other possible objectives for businesses, such as consideration of the interests of stakeholders."

Paragraph 3:

- The problem here is not the functioning of the market, but instead all the people not fully understanding the impact of their consumption: with full information, people's wants will be satisfied by the market,
- *Passage Example:* "Dumping toxic chemicals into public rivers... would be very damaging to the environment." If the pollution is common knowledge, and if the majority of people think it is bad, they will buy goods from an alternative green business, and the polluting company will go out of business.
- *Passage Example:* "A decision to transfer a call centre from the UK to India ... would make many employees redundant." If this is common knowledge the majority will switch their consumption to another business; indeed, we see BT had to relocate back to the UK to re-cooperate customers.

Paragraph 4:

- It may not be possible to fully inform customers, in which case the argument for regulation enforcing business responsibility is clear: external referees are needed to identify the good which is preferred for people and ensure they get the right one in the market.
- *Passage Example*: "Many borrowers stopped meeting their repayments..." Borrowers bought loans that were not fully understood by the players in the market and were too risky.

Conclusion

- Summarize the points:
  - Allocation is determined by people's demand: any interferences undermine people's wants.
  - Allocations may affect other stakeholders resulting in net-loss of welfare.
  - If all stakeholders (consumers, businesses and externally affected people) understand all the repercussions of production, wants will be satisfied directly in the market.
  - Full information for everyone is not feasible in some contexts.
- Now decide for yourself one way or another...
- Zoom out and say why the question really matters (just pick one to give your essay the 'wow-factor'):
  - Government Intervention is needed to ensure market interactions are net-beneficial to society, not just to the active buyers and sellers. CSR gives a voice to the wants of the silent victims of markets.

Passage 5

When discussing his famous character Rorschach, the antihero of 'Watchmen', Moore explains, 'I originally intended Rorschach to be a warning about the possible outcome of vigilante thinking. But an awful lot of comic readers felt his remorseless, frightening, psychotic toughness was his most appealing characteristic – not quite what I was going for.' Moore misunderstands his own hero's appeal within this quotation: it is not that Rorschach is willing to break little fingers to extract information, or that he is happy to use violence, that makes him laudable. The Comedian, another 'superhero' within the alternative world of Watchmen, is a thug who has won no great fan base; his remorselessness (killing a pregnant Vietnamese woman), frightening (attempt at rape), psychotic toughness (one only has to look at the panels of him shooting out into a crowd to witness this) is repulsive, not winning. This is because The Comedian has no purpose: he is a nihilist, and as a nihilist, denies any potential meaning to his fellow man, and so to the comic's reader. Everything to him is a 'joke', including his self, and consequently, his own death could be seen as just another gag.

Rorschach, on the other hand, does believe in something: he questions if his fight for justice 'is futile?' then instantly corrects himself, stating 'there is good and evil, and evil must be punished. Even in the face of Armageddon I shall not compromise in this.' Jacob Held, in his essay comparing Rorschach's motivation with Kantian ethics, put forward the postulation that 'perhaps our dignity is found in acting as if the world were just, even when it is clearly not.' Rorschach then causes pain in others not because he is a sadist, but because he feels the need to punish wrong and to uphold the good, and though he cannot make the world just, he can act according to his sense of justice - through the use of violence.

QUESTION

"There is no difference between Rorschach's and The Comedian's use of violence because wrong and right are subjective notions." Discuss with reference to the passage above.

Example Plan

Introduction:

- Define subjectivity: the notion that individuals can view the same thing and interpret it differently. Why does this matter in to the question? Because a superhero's interpretation of a situation will justify – or not – their use of violence in response to the situation.
- The key question: does the impossibility of defining what is right or wrong make a superhero uses violence to exact justice indistinguishable from another violent superhero (or villain)?

Paragraph 1:

- Subjectivity matters to differentiating between the two superheroes, because if Rorschach's vision of 'righting wrongs', as a vigilante for justice, depends on him *correctly identifying* wrongs that *everyone* thinks are wrongs (i.e. objective evils). Someone may disagree in his interpretation of something as wrong, and thus view Rorschach as a villain.
- *Passage Example*: The author says "Rorschach acts according to his *sense of justice*." His view of justice may not be shared with others.
- *Passage Example*: The author intended Rorschach to be a "warning about the potential outcome of vigilante thinking" but "misunderstands his own hero's appeal." The author and the audience differ in their interpretations.
- *[Current affairs Example*: People's polarized view of the American President: one person's hero is another's villain.]

Paragraph 2:

- Subjectivity does not matter because it is the *intention* to do right that justifies the use of violence; this *intention* is what differentiates the two superheroes. One superhero is trying to be a force for good, and therefore is laudable. The other superhero acts with no purpose.
- *Passage Example*: "The Comedian has no purpose: he is a nihilist."
- *Passage Example*: Rorschach acts "according to his sense of justice."

Paragraph 3:

- It is the *means*, not the intention that defines a superhero: if both resort to violence, then they cannot be differentiated in terms of what they do. Violence will always cause pain to someone, and often those affected will be innocents.
- *Passage Example:* The outcome of Rorschach's actions is pain and so is the outcome of The Comedian's.

Paragraph 4:

- What makes a superhero different from a villain is that despite the difficulty of identifying rights and wrongs, they are still willing to fight for something better than the status quo, which is unjust.
- *Passage Example*: Jacob Held's discussion of Kantian Ethics: in the face of inevitable evil, Rorschach fights for something better. This gives his acts and the world meaning.
- *Passage Example*: "The Comedian denies any potential meaning to his fellow man."

Conclusion

- Summarize the points:
    - One person's superhero is another man's villain: Rorschach and The Comedian cannot be distinguished.
    - It is a superhero's *intention* to do right that distinguished them from the villain.
    - Using violence means that the two superheroes cannot be distinguished, because they are defined by the means of exacting justice, not their intention.
    - It is a superhero's willingness to engage with an unjust and subjective world that differentiates him.
- Zoom out and say why the question of violence in the face of a subjective notion of justice really matters (there are so many applications of this – just pick one to give your essay the 'wow-factor') e.g. Use of torture by governments, Concept of a just war, Capital punishment

Intro
- Does it explain or just repeat?
- Does it set up the main body?
- Does it get to the point?

Main Body
- Are enough points being made? *[Breadth]*
- Are the points explained sufficiently? *[Depth]*
- Does the argument make sense? *[Strength]*

Conclusion
- Does it follow naturally from the main body?
- Does it consider both sides of the argument?
- Does it answer the original question?

Final Advice

✓ Always answer the question clearly – this is the key thing examiner look for in an essay.

✓ Analyse each argument made, justifying or dismissing with logical reasoning.

✓ Keep an eye on the time/space available – an incomplete essay may be taken as a sign of a candidate with poor organisational skills.
✓ Ensure each paragraph has a new theme that is clearly differentiated from the previous one (don't just use a new paragraph to break your text up)
✓ Leave yourself time to write a conclusion – however short – that tells your reader which side of the fence you're on
✓ Do plan your essay before you start writing even the introduction; don't be tempted to dive straight into it
✓ Use pre-existing knowledge when possible – examples and real world data can be a great way to strengthen an argument- but don't make up statistics!

✓ Present ideas in a neat, logical fashion (easier for an examiner to absorb).

✓ Complete some practice questions in advance, in order to best establish your personal approach to the paper (particularly timings, how you plan etc.).

✗ Attempt to answer a question that you don't fully understand, or ignore part of a question.
✗ Rush or attempt to use too many arguments – it is much better to have fewer, more substantial points.
✗ Attempt to be too clever, or present false knowledge to support an argument – a tutor may call out incorrect facts etc.
✗ Panic if you don't know the answer the examiner wants – there is no right answer, the essay is not a test of knowledge but a chance to display reasoning skill. Start by defining the words in the question to get your mind thinking about ways to approach it
✗ Leave an essay unfinished – if time/space is short, wrap up the essay early in order to provide a conclusive response to the question. If you've only got a couple of minutes left, summarize your remaining points in a short bullet point each; these bullets contain just the topic sentence and (optionally) a quote from the passage to illustrate your point

ANSWERS

Answer Key

Q	A	Q	A	Q	A	Q	A	Q	A	Q	A	Q	A
1	B	51	A	101	C	151	C	201	D & E	251	A	301	E
2	E	52	C & E	102	A	152	B	202	C	252	C	302	A
3	E	53	D	103	B	153	E	203	C	253	C	303	C
4	D	54	D	104	B	154	C	204	C	254	C	304	C
5	E	55	B	105	E	155	B	205	C	255	B	305	E
6	E	56	D	106	E	156	E	206	D	256	D	306	A
7	A	57	A	107	C	157	D	207	C	257	E	307	D
8	C	58	B	108	B	158	C	208	B	258	D	308	C
9	D	59	C	109	C	159	B	209	C	259	D	309	D
10	C	60	B	110	D	160	C	210	C	260	B	310	A
11	E & F	61	C	111	C	161	D	211	C	261	E	311	D
12	D	62	D	112	B	162	A	212	E	262	E	312	B
13	C	63	D	113	D	163	D	213	A	263	B	313	E
14	B	64	C	114	C	164	B	214	C	264	C	314	A
15	A	65	C	115	C	165	D	215	E	265	A	315	D
16	C	66	C	116	A	166	D	216	E	266	C	316	C
17	B	67	A	117	C	167	C	217	C	267	B	317	E
18	C	68	C	118	B	168	C	218	E	268	B		
19	D	69	A	119	E	169	C	219	E	269	B		
20	B & C	70	E	120	D	170	A	220	E	270	C		
21	B	71	C	121	C	171	B	221	B	271	C		
22	B & D	72	B	122	B	172	D	222	C	272	A		
23	D	73	B	123	E	173	B	223	B	273	C		
24	F	74	C	124	C	174	A	224	B	274	D		
25	D	75	B	125	C	175	D	225	C	275	C		
26	D	76	C	126	E	176	B	226	D	276	D		
27	B & D	77	E	127	B	177	C	227	C	277	A		
28	B	78	D	128	D	178	B	228	B	278	C		
29	C	79	C	129	C	179	D	229	A	279	B		
30	C	80	B	130	D	180	F	230	F	280	B		
31	D	81	E	131	E	181	C	231	D	281	A		
32	B	82	D	132	B	182	B	232	A	282	C		

33	C	**83**	A	**133**	C	**183**	A	233	B	**283**	C		
34	B	**84**	C	**134**	C	**184**	B	234	A	**284**	B		
35	E	**85**	A	**135**	D	**185**	A	235	F	**285**	A		
36	B	**86**	B	**136**	B	**186**	C	236	D	**286**	C		
37	D	**87**	C	**137**	E	**187**	C	237	A	**287**	E		
38	D	**88**	A	**138**	C	**188**	C	238	D	**288**	C		
39	C	**89**	B	**139**	A	**189**	C	239	D	**289**	C		
40	C	**90**	F	**140**	C	**190**	C	240	F	**290**	E		
41	B	**91**	C	**141**	C	**191**	D	241	B	**291**	C		
42	C	**92**	E	**142**	D	**192**	C	242	C	**292**	C		
43	C	**93**	B	**143**	A	**193**	D	243	B	**293**	A		
44	C	**94**	B	**144**	D	**194**	C	244	C	**294**	D		
45	D	**95**	C	**145**	D	**195**	C	245	A	**295**	C		
46	C	**96**	C	**146**	C	**196**	C	246	A	**296**	B		
47	E	**97**	C	**147**	C	**197**	C	247	C	**297**	A		
48	D	**98**	F	**148**	C	**198**	E	248	B	**298**	C		
49	C	**99**	C	**149**	E	**199**	C	249	D	**299**	C		
50	A	**100**	A	**150**	D	**200**	C	250	C	**300**	E		

Worked Answers

Question 1: B
By making a grid and filling in the relevant information the days Dr James works can be deduced:

	Sunday	Monday	Tuesday	Wednesday	Thursday	Friday	Saturday
Dr Evans	X	√	X	X	√	√	√
Dr James	X	√	√	√	√	X	√
Dr Luca	X	X	√	√	X	√	√

- No onc works Sunday.
- All work Saturday.
- Dr Evans works Mondays and Fridays.
- Dr Luca cannot work Monday or Thursday.
- So, Dr James works Monday.
- And, Dr Evans and Dr James must work Thursday.
- Dr Evans cannot work 4 days consecutively so he cannot work Wednesday.
- Which means Dr James and Luca must work Wednesday.
- (mentioned earlier in the question) Dr Evans only works 4 days, so cannot work Tuesday.
- Which means Dr James and Luca work Tuesday.
- Dr James cannot work 5 days consecutively so cannot work Friday.
- Which means Dr Luca must work Friday.

Question 2: E
Working algebraically, using the call out rate as C, and rate per mile as M.
So, C + 4m = 11
C + 5m = 13
Hence; (C + 5m) – (C + 4m) = £13 - £11
M = £2
Substituting this back into C + 4m = 11
C + (4 x 2) – 11
Hence, C = £3
Thus a ride of 9 mile will cost £3 + (9 x £2) = £21.

Question 3: E
Use the information to create a Venn diagram.
We don't know the exact position of both Trolls and Elves, so **A** and **D** are true. Goblins are mythical but not magical, so **C** is true. Gnomes are neither so **B** is true. But **E** is not true.

Question 4: D
The best method may be work backwards from 7pm. The packing (15 minutes) of all 100 tiles must have started by 6:45pm, hence the cooling (20 minutes) of the last 50 tiles started by 6:25pm, and the heating (45 minutes) by 5:40pm. The first 50 heating (45 minutes) must have started by 4:35pm, and cooling (20 minutes) by 5:20pm. The decoration (50 minutes) of the second 50 can occur anytime during 4:35pm- 5:40pm as this is when the first 50 are heating and cooling in the kiln, and so does not add time. The first 50 take 50 minutes to decorate and so must be started by 3:45pm.

Question 5: E
Speed = distance/time. Hence for the faster, pain impulse the speed is 1m/ 0.001 seconds. Hence the speed of the pain impulse is 1000 metres per second. The normal touch impulse is half this speed and so is 500 metres per second.

Question 6: E

Using the months of the year, Melissa could be born in March or May, Jack in June or July and Alina in April or August. With the information that Melissa and Jack's birthdays are 3 months apart the only possible combination is March and June. Hence Alina must be born in August, which means it is another 7 months until Melissa's birthday in March.

Question 7: A

PC Bryan cannot work with PC Adams because they have already worked together for 7 days in a row, so **C** is incorrect. **B** is incorrect because if PC Dirk worked with PC Bryan that would leave PC Adams with PC Carter who does not want to work with him. PC Carter can work with PC Bryan.

Question 8: C

Paying for my next 5 appointments will cost £50 per appointment before accounting for the 10% reduction, hence the cost counting the deduction is £45 per appointment. So the total for 4 appointments = 5 x £45 = £225 for the hair. Then add £15 for the first manicure and £10 x 2 for the subsequent manicures using the same bottle of polish bringing an overall total of £260.

Question 9: D

Elena is married to Alex or David, but we are told that Bertha is married to David and so Alex must be married to Elena. Hence David, Bertha, Elena and Alex are the four adults. Bertha and David's child is Gemma. So Charlie and Frankie must be Alex and Elena's two children. Leaving only options **A** or **D** as possibilities. Only Frankie and Gemma are girls so Charlie must be a boy.

Question 10: C

Using, x (minutes) as the, unknown amount of time, the second student took to examine, we can plot the time taken with the information provided thus:

	1st student		2nd student		3rd student
1st examination:	4x	1	2x	1	2x
			Break: 8 minutes		
1st examination:	x	1	x	1	x

Hence the total time taken, 45minutes (14:30-15:15)

Is represented by, $4x + 2x + 2x + x + x + x + 1 + 1 + 8 + 1 + 1$

$45 \text{ minutes} = 11x \text{ (minutes)} + 12 \text{ minutes}$

$33 \text{ minutes} = 11x \text{ (minutes)}$

Hence, x = 3 minutes, so the amount of time the second student took the first time, 2x, is 6 minutes.

Question 11: E & F

To work out the amount of change is the sum £5 - (2 x £1.65), which = £3.30. Logically we can then work out that the 3 coins in the change that are the same must be 1p as no other 3 coin combination can yield £1.70 when made up with 5 more coins. Thus we know that 3 of the coins are 1p, 1p & 1p. We can then deduce that there must also have been 2p and 5p coins in the change as £1.70 is divisible by ten. The only way then to make up the remaining £1.60 in 3 different coins is to have £1, 50p and 10p, Hence the change in coins is 1p, 1p, 1p, 2p, 5p, 10p, 50p and £1. So the two coins not given in change are £2 and 20p.

Question 12: D

If we express the speed of each train as W ms^{-1}. Then the relative speed of the two trains is 2W ms^{-1}.

Using Speed=distance/time: 2W = (140 + 140)/ 14.

Thus, 2W = 20, and W = 10. Thus, the speed of each train is 10 ms^{-1}.

To convert from metres to kilometres, divide by 1,000. To convert from seconds to hours, divide by 3,600.

Therefore, the conversion factor is to divide by 1,000/3,600 = 10/36 = 5/18

Thus, to convert from ms^{-1} to kmph, multiply by 18/5. Therefore, the final speed of the train is 18/5 x 10 = 36km/hr.

Question 13: C
Taking the day to be 24 hours long, this means the first tap fills 1/6 of the pool in an hour, the second 1/48, the third $\frac{1}{72}$ and the fourth $\frac{1}{96}$.

Taking 288 as the lowest common denominator, this gives: $\frac{48}{288} + \frac{6}{288} + \frac{4}{288} + \frac{3}{288}$ which $= \frac{61}{288}$ full in one hour. Hence the pool will be $\frac{244}{288}$ full in 4 hours.

The pool fills by approximately $\frac{15}{288}$ every 15 minutes.

Thus, in 4 Hours 15: $\frac{244 + 15}{288} = \frac{249}{288}$

Thus, in 4 Hours 30: $\frac{244 + 30}{288} = \frac{274}{288}$

Thus, in 4 Hours 45: $\frac{244 + 45}{288} = \frac{289}{288}$

Question 14: B
Every day up until day 28 the ant gains a net distance of 1cm, so at the end of day 27 the ant is at 27cm height and therefore only 1cm below the top. On day 28 the 3cm the ant climbs in the day is enough to take it to the top of the ditch and so it is able to climb out.

Question 15: A
To solve this question three different sums are needed to use the information given to deduce the costs of the various items. With the information that 30 oranges cost £12, £12/30 = 40p per orange with the 20% discount, hence oranges must cost 50p at full price. With the information that 5 sausages and 10 oranges cost £8.50, we know that the oranges at a 10% discount account for 10 x 45p = £4.50 so 5 undiscounted sausages cost £4 so each full price sausage is £4/5 = 80p. Finally we know that 10 sausages and 10 apples cost £9, at 10% discount the sausages cost 72p each thus accounting for 10 x 72p = £7.20 of the £9, hence the 10 apples at a 10% discount must cost £1.80, so each apple costs 18p at 10% discount. So an apple is 20p full price. Now to add up the final total: 2 oranges + 13 sausages + 2 apples = (2 x 50p) + (13 x 72p) + (12 x 18p) = £12.52.

Question 16: C
If we take the number of haircuts per year to be x, the information we have can be shown:

Membership	Annual Fee	Cost per cut	Total Yearly cost
None	None	£60	60x
VIP	£125	£50	£125 + 50x
Executive VIP	£200	£45	£200 + 45x

As we know that changing to either membership option would cost the same for the year, we can express the cost for the year, y as;
VIP: y = £125 + 50x
Executive VIP: y = £200 + 45x
Therefore: £125 + 50x = £200 + 45x
Simplified 5x = £75, therefore the number of haircuts a year, x is 15.
Substituting in x, we can therefore work out:

Membership	Annual Fee	Cost per cut	Total Yearly cost
None	None	£60	£900

VIP	£125	£50	£875
Executive VIP	£200	£45	£875

Hence the amount saved by buying membership is £25.

Question 17: B

All thieves are criminals. So the circle must be fully inside the square, we are told judges cannot be criminals so the star must be completely separate from the other two.

Question 18: C

We are told that March and May have the same last number, which must be either 3 or 13. Taking the information from the question that one of the factors is related to the letters of the month names, we can interpret that 13 represents the M which starts both March and May. Therefore we know the rule is that the last number is the position of the starting letter. Knowing that there is another factor about the letters of the month that controls the code we can work out that one of the number may code for the number of letters. Which in March would be 5, which is the second letter, so we have the rule of the 2nd number. Finally through observation we may note that the first number codes for the months' relative position in the year. Hence the code of April will be 4, (for its position), 5 (for the number of letters in the name) and 1 for the position of the starting letter 'A') and so 451 is the code.

Question 19: D

If b is the number of years older than 5, and a the number of A*s, the money given to the children can be expressed:

£5 + £3b + £10a

Hence for Josie £5 + (£3 x 11) + (£10 x 9) = £128

We know that Carson receives £44 less yearly, and his b value is 13, so his amount can be expressed:

£5 + (£3 x 13) + (£10a) = £84

Simplified: £44 + £10a = £84

I.e. £10a = £40,

So Carson's 'a' value, i.e. his number of A*s is 4, so the difference between Josie and Carson is 5.

Question 20: B & C

Using the information to make a diagram:

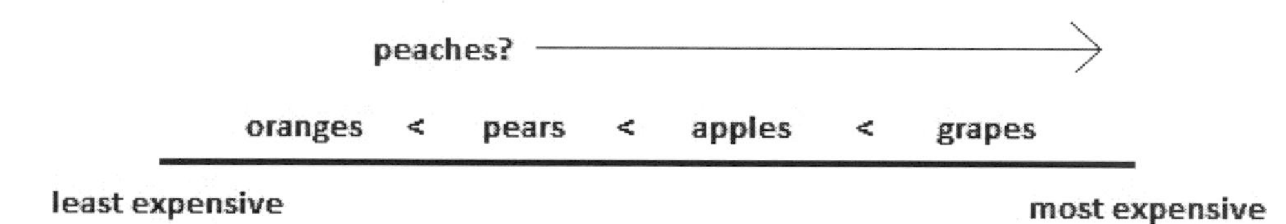

Hence **A** is incorrect. **D** and **E** may be true but we do not have enough information to say for sure. **B** is correct as we know peaches are more expensive than oranges but not about their price relative to pears. Equally we know **C** to be true as grapes are more expensive than apples so they must be more expensive than pears.

Question 21: B

It's easy to assume all the cuts should be in the vertical plane as a cake is usually sliced, however there is a way to achieve this with fewer cuts. Only three cutting motions are needed. **Start by cutting in the horizontal plane** through the centre of the cake to divide the top half from the bottom half. Then slice in the vertical plane into quarters to give 8 equally sized pieces with just three cuts.

Question 22: B & D

After the changes have been made, at 12 PM (GMT +1):

- Russell thinks it is 11 AM
- Tom thinks it is 12 PM
- Mark thinks it is 1 PM

Thus, in current GMT+1 time zone, Mark will arrive an hour early at 11 AM, Russell an hour late at 1 PM and Tom on time at 12 PM. There is therefore a two hour difference between the first and last arrival. For options E and F, be careful: the time zone listed is **NOT** GMT +1 that everyone else is working in. 1PM in GMT +3 =

11pm GMT +1 (the summer time zone just entered) so that is Mark's actual arrival time; 12pm GMT +0 is the old time zone that Russell didn't change out of so that is Russell's correct arrival time.

Question 23: D
Using Bella's statements, as she must contradicted herself with her two statements, as one of them must be true, we know that it was definitely either Charlotte or Edward. Looking to the other statements, e.g. Darcy's we know that it was either Charlotte or Bella, as only one of the two statements saying it was both of them can have been a lie. Hence it must have been Charlotte.

Question 24: F
The only way to measure 0.1 litres or 100ml, is to fill the 300ml beaker, pour into the half litre/ 500 ml beaker, fill the 300ml again and pour (200ml) into the 500ml, which will make it full, leaving 100ml left in the 300ml beaker. The process requires 600ml of solution to fill the 300ml beaker twice.

Question 25: D
If you know how many houses there are on the street it is possible to work out the average, which then you can round up and down and to find the sequence of number, e.g. if you know there are 6 houses in the street 870/ 6 = 145. Which is not a house number because they are even so going up and down one even number consequentially one discovers that the numbers are 146, 144, 148, 150, 142 and 140. But it is not possible to determine Francis' house number without knowing its relative position i.e. highest, 3rd highest, lowest etc.

Question 26: D
Expressed through time:

Event	People Present
There were 20 people exercising in the cardio room	**20**
Four people were about to leave	**20**
A doctor was on the machine beside him (one of the original 20)	**20**
Emerging from his office one of the personal trainers called an ambulance.	**21**
Half of the people who were leaving, left (-2)	**19**
Eight people came into the room to hear the man being pronounced dead. (+8)	**27**
the two paramedics arrived, (+2)	**29**
the man was pronounced dead (-1)	**28**

Question 27: B & D

Blood loss can be described as 0.2 L/min.
For the man:
8 litres – 40% (3.2 L) = 4.8 L When he collapses, taking 16 minutes (3.2 / 0.2 = 16)
For the woman:
7 litres – 40% (2.8L) = 4.2: when she collapses, taking 14 minutes (2.8 / 0.2 = 14)
Hence the woman collapses 2 minutes before the man so **B** is correct, and **A** is incorrect. The total blood loss is 3.2L + 2.8L which = 6L so **C** is incorrect. The man's blood loss is 3.2L when he collapses so **E** is incorrect. The woman has a remaining blood volume of 4.2L when she collapses so **D** is correct. Blood loss is 0.2 L/min, which equates to 5 minutes per litre, which is 10 minutes per 2 litres not 12 L, so **F** is incorrect.

Question 28: B
Work out the times taken by each girl – (distance/pace) x 60 (converts to minutes) + lag time to start
Jenny: (13/8) x 60 = 97.5 minutes
Helen: (13/10) x 60 + 15 = 93 minutes
Rachel (13/11) x 60 + 25 = 95.9 minutes

Question 29: C
Work through each statement and the true figures.

A. Overlap of pain and flu-like symptoms must be at least 4% (56+48-100). 4% of 150: 0.04 x 150=6
B. 30% high blood pressure and 20% diabetes, so max percentage with both must be 20%. 20% of 150: 0.2*150 = 30

C. Total number of patients – patients with flu-like symptoms – patients with high blood pressure. Assume different populations to get min number without either. 150 – (0.56 x 150) – (0.3 x 150) = 21
D. This is an obvious trap that you might fall into if you added up the percentages and noted that the total was >100%. However, this isn't a problem as patients can discussed two problems.

Question 30: C

This is easiest to work out if you give all products an original price, I have used £100. You can then work out the higher price, and the subsequent sale price, and thus the discount from the original £100 price. As the price increases and decreases are in percentages, they will be the same for all items regardless of the price so it does not matter what the initial figure you start with is.
Marked up price: 100 x 1.15 = £115
Sale price: 115 x 0.75 = £86.25
Percentage reduction from initial price is 100 – 86.25 = 13.75%

Question 31: D

The recipe states 2 eggs makes 12 pancakes, therefore each egg makes 6 pancakes, so the number Steve must make should be a multiple of 6 to ensure he uses a whole egg.
Steve requires a minimum of 15 x 3 = 45 pancakes. To ensure use of whole eggs, this should be increased to 48 pancakes.
The original recipe is for 12 pancakes, therefore to make 48 pancakes, require 4x recipe (48/12).
Therefore quantities: 8 eggs, 400g plain flour and 1200 ml milk.

Question 32: B

Work through the question backwards.
In 6 litres of diluted bleach, there are 4.8 litres of water and 1.2 litres of partially diluted bleach.
In the 1.2 litres of partially diluted bleach, there is 9 parts water to one part original warehouse bleach. Remember that a ratio of 1:9 means 1/10 bleach and 9/10 water. Therefore working through, there is 120ml of warehouse bleach needed.

Question 33: C

We know that Charles is born in 2002, therefore in 2010 he must be 8. There are 3 years between Charles and Adam, and Charles is the middle grandchild. As Bertie is older than Adam, Adam must be younger than Charles so Adam must be 5 in 2010. In 2010, if Adam is 5, Bertie must be 10 (states he is double the age of Adam).
The question asks for ages in 2015: Adam = 10, Bertie = 15, Charles = 13

Question 34: B

Make the statements into algebraic equations and then solve them as you would simultaneous equations. Let a denote the flat fixed rate for hire, and b the price per half hour.
Cost = a + b(time in mins/30)
Peter: a + 6b (6 half hours) = 14.50 (equation 1)
Kevin: 2a + 18b = 41, or this can be simplified to give cost per kayak, a + 9b = 20.5 (equation 2)
If you subtract equation 1 from equation 2:
3b = 6, therefore b = 2

Substitute b into either equation to calculate a, using equation 1, a + 12 = 14.50, therefore a = 2.50
Finally use these values to work out the cost for 2 hours:
2.50 (flat fee) + 4 x 2 (4half hours x cost/half hour) = £10.50

Question 35: E

It is most helpful to write out all the numbers from 0 – 9 in digital format to most easily see which light elements are used for each number. You can then cross out any numbers which don't use all the lights from the digit 7.

Go through the digits methodically and you can cross out: 1, 2, 4, 5, and 6. These numbers don't contain all three bars from the digit 7.

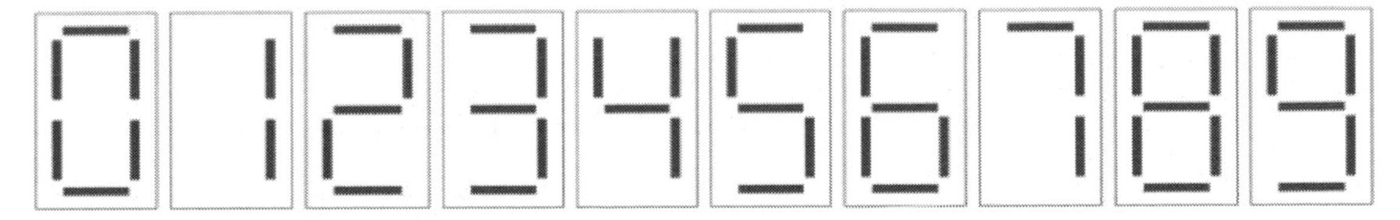

Question 36: B
In this question it is worth remembering it will take more people a shorter amount of time.
Work out how many man hours it takes to build the house. Days x hours x builders
12 x 7 x 4 = 336 hours
Work out how many hours it will take the 7man workforce: 336/7 = 48 hours
Convert to 8 hour days: 48/8 = 6 days

Question 37: D
By far the easiest way to do these type of questions is to draw a Venn diagram (use question marks if you are unsure about the exact position):

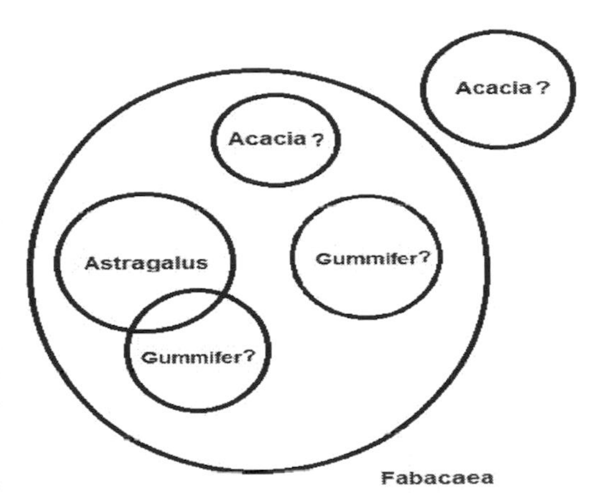

Now, it is a case of going through each statement:

A. Incorrect - Acacia may be fabacaea. Acacia are not astragalus, but does not logically follow that they therefore can't be fabacaea.
B. Incorrect – astragalus and gummifer are not necessarily separate within fabacaea.
C. Incorrect – the statement is not reversible so the fact that all astragalus and gummifer are fabacaea does not mean all facacaea are gummifer and/or astragalus. E.g. Fabacaea could be acacia.
D. Correct
E. Incorrect – Whilst some acacia could be gummifer, there is no certainty that they are.

Question 38: D
Area of a trapezium = (a+b)/2 x h
Area of cushion = (30+50)/2 x 50 = 2000cm^2
Since each width of fabric is 1m wide, both sides of one cushion can fit into one width. The required length is therefore 75cm x 4 = 3m with a cost of 3 x £10 = £30.
Cost of seamstress = £25 x 4 = £100
Total cost is £130

Question 39: C
There are 30 days in September, so Lisa will buy 30 coffees.
In Milk, every 10th coffee is free, so Lisa will pay for 27 coffees at 2.40 = £64.80
In Beans, Lisa gets 20 points each day and needs 220 points to get a free coffee, which is 11 days, with 5 points left over. Therefore, in 30 days she will get 2 free coffees. The cost for 28 coffees at 2.15 is £60.20
Beans is cheaper, and the difference is £64.80 - £60.20 = £4.60.

Question 40: C
Work backwards and take note of how often each bus comes.
Must get off 220 bus at 10.57 latest. Can get 10.40 bus therefore (arrive at 10.54).
Latest can get on 283 bus is 10.15 as to make the 220 bus connection. 283 comes every 10mins (question doesn't state at what points past the hour), so Paula should be at the bus stop at 10.06 to ensure a bus comes by 10.15 at the latest. If the bus comes every 10mins, even if a bus comes at 10.05 which Paula will miss, the next bus will come at 10.15 and therefore she will still be on time.
Therefore Paula must leave at 10.01

Question 41: B

You are working out the time taken to reach the same distance (D). Make sure to take into account changing speeds of train A, and that train B leaves 20 minutes earlier.

$Speed = \frac{distance}{time}$

Make sure you keep the answers consistent in the time units you are using, the worked answer is all in minutes (hence the need to multiply by 60).

Train A: time for first $20km = \frac{20}{100} x\, 60 = 12\ minutes$

So the distance where it equals B is $12 + (\frac{D-20}{150})\, x\, 60$

You need to use D-20 to account for the fact you have already calculated the time at the slower speed for the first 20km

Train B: $(\frac{D}{90})\, x\, 60 - 20$

Make the equations equal each other as they describe the same time and distance, and solve.

Simplifies to $32 + \frac{2D}{5} - 8 = \frac{2D}{3}$ so $D = 90km$

Train B will take 60 minutes to travel 90 km and train A will take 40 minutes (but as it leaves 20 minutes later, this will be point at which it passes).

Question 42: C

Work out the annual cost of local gym: 12 x 15 = £180

Upfront cost + class costs of university gym must therefore be >£180.

Subtract upfront cost to find number of classes: 180 – 35 = £145

Divide by cost per class (£3) to find number of classes: 145/3 = 48 1/3

48 1/3 classes would make the two gyms the same price, so for the local gym to be cheaper, you would need to attend 49 classes.

Question 43: C

A is definitely true, since the question states that all herbal drugs are not medicines. **B** is also definitely true as all antibiotics are medicines which are all drugs. **C** is definitely false, because all antibiotics are medicine, yet no herbal drugs are medicines. **D** is true as all antibiotics are medicines.

Question 44: C

Answer **A** cannot be reliably concluded, because from the information given a non-"Fast" train could stop at Newark, but not at Northallerton or Durham. We have no information on whether *all* trains stopping at Newark also stop at Northallerton.

Answer **B** is not correct because 8 is the *average* number of trains that stop at Northallerton. It is possible that on some days more than 16 trains run, and more than 8 will thus stop at Northallerton. Answer **D** is incorrect because it is mentioned that *all* trains stopping at Northallerton also stop at Durham, giving a total 6 stops as a minimum for a train stopping at Northallerton (the others being the 4 stops which *all* trains stop at).

Answer **E** is incorrect for a similar reason to **A**. We have no information on whether all trains stopping at Newark also stop at Northallerton, so cannot determine that they must also stop at Durham.

Answer **C** is correct because "Fast" trains make less than 5 stops. Since all trains already stop at 4 stops (Peterborough, York, Darlington and Newcastle), they cannot then stop at Durham, as this would give 5 stops.

Question 45: D
From the information we are given, we can compose the following image of how these towns are located (not to scale, but shows the direction of each town with respect to the others):

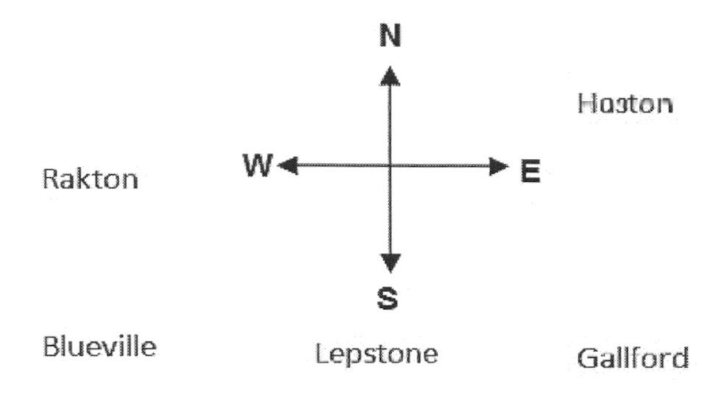

From this "map", we can see that all statements apart from **D** are true. Statement **D** is definitely *not true*, since Blueville is south west of Haston it cannot be East of Haston.

Question 46: C
We are told that in order to form a government, a party (or coalition) must have *over* 50% of seats. Thus, they must have at least 50% of the total seats plus 1, which is 301 seats.
We are told that we are looking for the *minimum* number of seats the greens can have in order to form a coalition with red and orange. Thus, we are seeking for Red and Orange to have the *maximum* number of seats possible, under the criteria given.

Thus we can calculate as follows:
- No party has over 45% of seats, so the maximum that the Red party can have is 45%, which is 270 seats.
- No party except for red and blue has won more than 4% of seats. We are told that the green party won the 4th highest number of seats, so it is possible that the Orange party won the 3rd highest.
- Thus, the maximum number of seats the orange party can have won is 4% of the total, which is 24 seats.
- Thus, the maximum possible combined total of the Red and Orange party's seats won is 294.

Thus, in order to achieve a total of 301 seats in a Red-Orange-Green coalition, the Green party have to have won at least 7 seats.

Question 47: E
Expressing the amount each child receives:

Youngest	M
2nd youngest	$M + D$
3rd youngest/ 3rd oldest	$M + 2D$
4th youngest/ 2nd oldest	$M + 3D$
Oldest	$M + 4D$

Question 48: D
The total amount of money received;
£100, = $M + M + D + M + 2D + M + 3D + M + 4D$
Simplified, thus is:
£100 = $5M + 10D$

Question 49: C
The two youngest are expressed as M and $M + D$. Simplified as $2M + D$.
The three oldest are expressed as $M + 2D$, $M + 3D$ and $M + 4D$, Simplified as $3M + 9D$
Hence 7 times the two youngest together is expressed $7(2M + D)$, so altogether the Answer is $7(2M + D) = 3M + 9D$.

Question 50: A
To work this out, simplify the two equations:

$7(2M + D) = 3M + 9D$

$14M + 7D = 3M + 9D$

$11M = 2D$

$M = \frac{2D}{11}$

Question 51: A
Substitute M into the equation £ $100 = 5M + 10D$

$5\left(\frac{2D}{11}\right) + 10D = £100$

$\frac{10D}{11} + 10D = \frac{10D}{11} + \frac{110D}{11} = \frac{120D}{11}$

Question 52: C & E
The easiest way to work this out is using a table. With the information we know:

1st		Madeira
2nd		
3rd	Jaya	
4th		

Ellen made carrot cake and it was not last. It now cannot be 1st or 3rd as these places are taken so it must be second:

1st		Madeira
2nd	Ellen	Carrot cake
3rd	Jaya	
4th		

Aleena's was better than the tiramisu, so she can't have come last, therefore Aleena must have placed first

1st	Aleena	Madeira
2nd	Ellen	Carrot cake
3rd	Jaya	
4th		

And the girl who made the Victoria sponge was better than Veronica:

1st	Aleena	Madeira
2nd	Ellen	Carrot cake
3rd	Jaya	Victoria Sponge
4th	Veronica	Tiramisu

Question 53: D
The information given can be expressed to show the results that the teams must have had to make their points total.

Team	Points	Game Results			
Celtic Changers	2	L	L	D	D
Eire Lions	?	?	?	?	?

Nordic Nesters	8	W	W	D	D
Sorten Swipers	5	W	D	D	L
Whistling Winners	1	D	L	L	L

The results so far total 3 wins, 6 losses and 7 draws. Since, the number of draws must be even, there must have been another draw, So we know one of the Eire Lions results is a draw.
The difference between wins (3) and losses (6) is 3. Thus, there must be another 3 wins to account for this difference. So the Eire Lions results must be 3 wins and 1 draw. Thus, they scored 3 x 3 + 1 = 10.

Question 54: D

Remember to consider the gender of each person. Then draw a quick diagram to show the given information you can see that only D is correct.

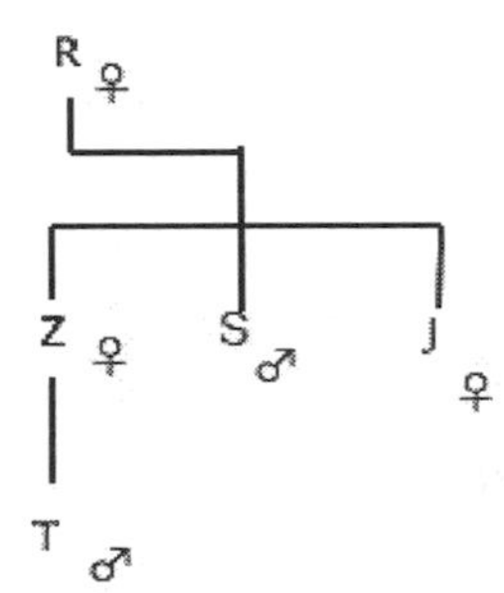

Question 55: B

After the first round; he knocks off 8 bottles to leave 8 left on the shelf. He then puts back 4 bottles. There are therefore 12 left on the shelf. After the second round, he has hit 3 bottles and damages 6 bottles in total, and an additional 2 at the end. He then puts up 2 new bottles to leave 12 – 8 + 2 = 6 bottles left on the shelf. After the final round, John knocks off 3 bottles from the shelf to leave 3 bottles standing.

Question 56: D

Based on the information we have we can plot the travel times below. Change over times are in a smaller font.

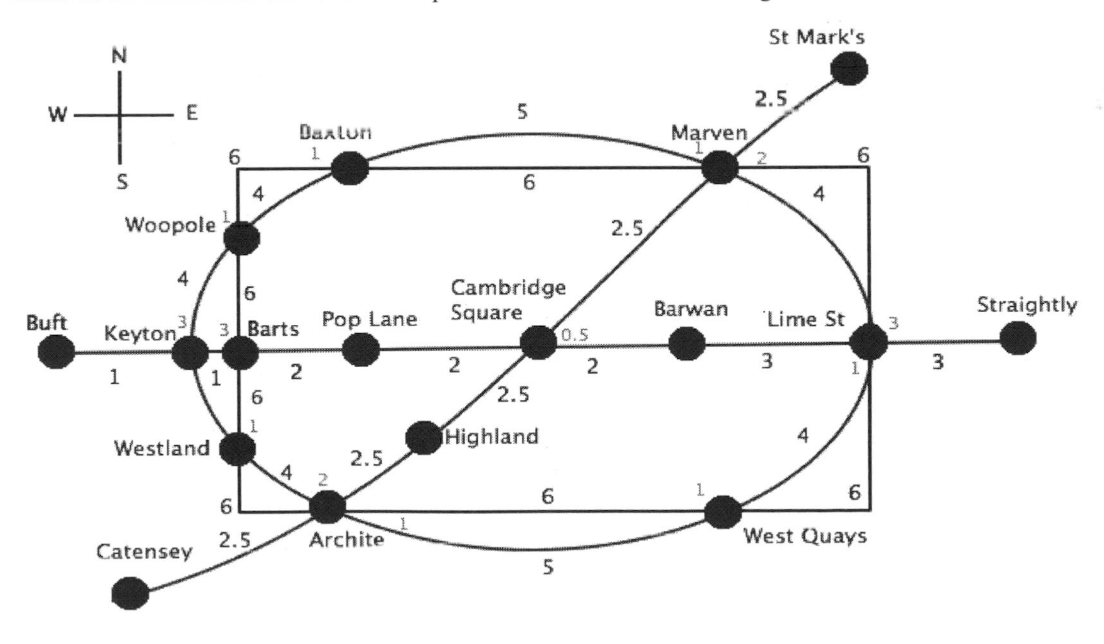

Hence on the St Mark's line, St Mark's to Archite takes 4 x 2.5 minutes = 10 minutes.

Question 57: A

Going from stop to stop on the Straightly line end Buft to Straightly would take 14 minutes, but we are told earlier on there is an express train that goes end to end and only takes 6.

Question 58: B

The quickest route from Baxton to Pop Lane is via Marven and Cambridge Square, which takes 5 + 2 + 2.5 + 0.5 + 2 = 12 minutes. Baxton to Pop Lane via Barts would take 4 + 1 + 6 + 3 + 2 = 16 minutes, which is longer so **E** is incorrect. Other options include times failing to take account of, or incorrectly adding changeover times, and so are incorrect.

Question 59: C

From Cambridge Square:

- Catensey is (2.5 x 3 =) 7.5 minutes away.
- Woopole, is (4 + 3 + 1 +2 + 2 =) 12 minutes.
- Buft is (1 + 1 + 2 + 2 =) 6 minutes.
- Westland is (4 + 2 + 2.5 + 2.5 =) 11 minutes.

Question 60: B

With the new delay information we can plot the travel times as before, adjusted for the delays. Plus a 5 minute delay on the platforms when waiting on any platform for a train.

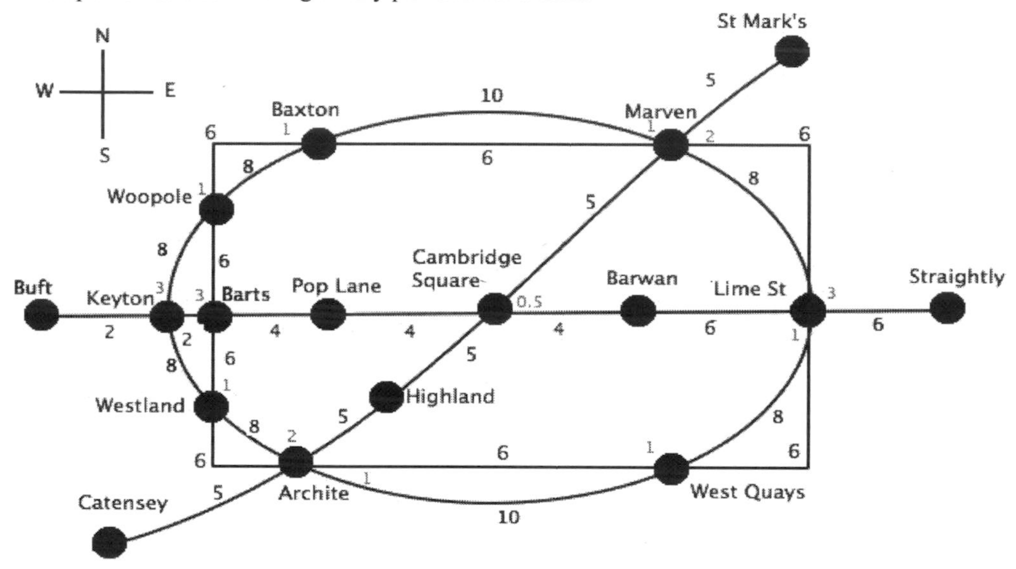

The quickest way from Westland to Marven now uses the non-delayed reliable rectangle line. Four stops on the rectangle line take 6 mins each so 24 minutes in total on the train. Add to this the additional 5 minutes platform waiting time to give a total journey time of 29 minutes.

Question 61: C

- Baxton to Archite via Barts using only the Rectangle line takes (5 + 6 +6+ 6 +6=) 29 minutes.
- Baxton to Woopole on the Rectangle line, then Oval to Archite via Keyton takes (5 + 6 + 1 + 5 + 8 + 8 + 8 =) 41 minutes
- Baxton to Archite on the Oval line only takes (5 + (8 x 4) =) 37 minutes
- Baxton to Woopole on the Oval line, then Rectangle to Archite via Barts takes (5 + 8 + 1 + 5 + 6 + 6 + 6 =) 37 minutes
- As the bus takes 27-31 minutes, it is not possible to tell from between the options which will be slower/ quicker so option **C** is the right answer.

Question 62: D

Remember the 5-minute platform wait. We are not told that the St Mark's express train from end to end is no longer running so we must assume that it is, which takes 5 minutes (plus the wait at St Mark's to go to Catensey).

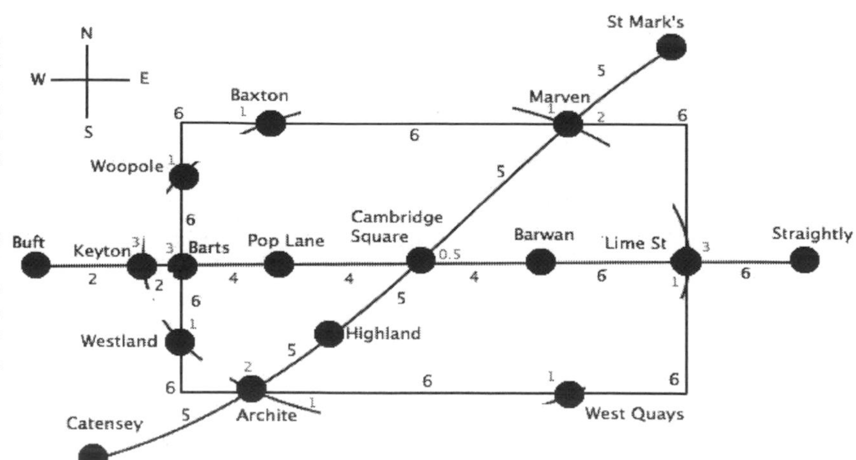

Then, there is a 5 minute wait at Catensey to Archite, and a 2 + 5 minute changeover at Archite onto the Rectangle line which then takes 6 minutes to West Quays. 5 + 5 + 5 + 5 + 2 + 5 + 6 = 33 minutes. Via

Lime St the journey takes 5 + 5 + 5+ 2 + 5 + 6+ 6 = 30 minutes.

Question 63: D

From the information:

- "Simon's horse wore number 1."
- "..the horse that wore 3, which was wearing red.."
- "the horse wearing blue wore number 4."

We can plot the information below:

Place	Owner	Number	Colours
	Simon	1	
		2	
		3	Red
		4	Blue

In addition: "The horse wearing green; Celia's, came second"

Which means Celia's horse must have worn number two because it cannot have worn number 1 because that is Simon's horse. Also it cannot have worn number three or four because they wore red and blue respectively. So we can plot this further deduction:

Place	Owner	Number	Colours
	Simon	1	
2nd	Celia	2	Green
		3	Red
		4	Blue

We also know that

- "Arthur's horse beat Simon's horse"
- "Celia's horse beat the horse that wore number 1." i.e. Simon's

We know Celia's horse came second, and that both Celia's and Arthur's horses beat Simon's. This means that Simon's horse must have come last. So;

Place	Owner	Number	Colours
4th	Simon	1	
2nd	Celia	2	Green
		3	Red
		4	Blue

And knowing that:

- "Only one horse wore the same number as the position it finished in."

The horses wearing numbers 3 and 4 must have placed 1st and 3rd respectively. Hence:

Place	Owner	Number	Colours
4th	Simon	1	
2nd	Celia	2	Green
1st		3	Red
3rd		4	Blue

"Lila's horse wasn't painted yellow nor blue"
So Lila's must have been red, and Simon's yellow. Leaving the only option for Arthur's to be blue. So we now know:

Place	Owner	Number	Colours
4th	Simon	1	Yellow
2nd	Celia	2	Green
1st	Lila	3	Red
3rd	Arthur	4	Blue

Question 64: C
Year 1 – 40 x 1.2 = 48
Year 2 – 48 x 1.2= 57.6
Year 3 – 57.6 x 1.1= 63.36
Year 4 – 63.36 x 1.1 = 69.696.

Question 65: C
To minimise the total cost to the company, they want the wage bills for each site to be less than £200,000. Working this out involves some trial and error; you can speed this up by splitting employees who earn similar amounts between the sites e.g. Nicola and John as they are the top two earners.
Nicola + Daniel + Luke = £ 198,500 and John + Emma + Victoria = £ 199,150

Question 66: C
Remember that pick up and drop off stops may be the same stop, therefore the minimum number of stops the bus had to make was 7. This would take 7 x 1.5 = 10.5 minutes.
Therefore the total journey time = 24 + 10.5 = 34.5 minutes.

Question 67: A
The best method here is to work backwards. We know the potatoes have to be served immediately, so they should be finished roasting at 4pm, so they should start roasting 50 minutes prior to that, at 3:10. We also know they have to be roasted immediately after boiling, so they should be prepared by 3:05, in order to boil in time. She should therefore start preparing them no later than 2:47, though she could prepare them earlier.
The chicken needs to be cooked by 3:55 to give it time to stand, so it should begin roasting at 2:40, and Sally should begin to prepare it no later than 2:25.
You can construct a rough timeline:

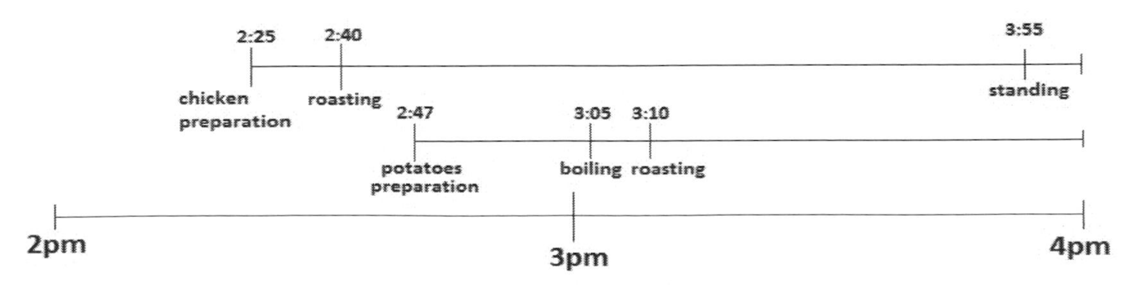

We can see from this timeline that from 2:40 onwards, there will be no long enough period of time in which there is a free space in the cooker for the vegetables to be boiled. They therefore must be finished cooking at 3:05. The latest time prior to this that Sally has time to prepare them (5 minutes) is at 2:40, between preparing the chicken and the potatoes. She should therefore begin preparing the vegetables at 2:42, then begin boiling at 2:47, so they can be finished cooking by 2:55, in time for the potatoes to boil at 3:05.
Chicken: 2:25
Potatoes: 2:47
Vegetables: 2:42

Question 68: C

The quickest way to do this is via trial and error. However, for the sake of completion: let each child's age be denoted by the letter of their name, and form an equation for their total age:

$P + J + A + R = 80$

The age of each child can be written in terms of Paul's age.

P = 2J, therefore $J = \frac{P}{2}$

$A = \frac{P+J}{2}$

Now substitute in $J = \frac{P}{2}$ to get in terms of P only: $A = \frac{P+\frac{P}{2}}{2} = \frac{P}{2} + \frac{P}{4} = \frac{3P}{4}$

$R = P + 2$

Thus: $P + \frac{P}{2} + \frac{3P}{4} + P(+ 2) = 80$

Simplify to give: $\frac{13P}{4} = 78$

$13P = 312$. Thus, $P = 24$

Substitute P = 24 into the equations for the other children to get: J = 12, A = 18, R = 26

Question 69: A

The total number of buttons is 71 + 86 + 83 = 240. The total number of suitable buttons is 22 + 8 = 30. Thus, she will have to remove a maximum of 210 buttons in order to guarantee picking a suitable button on the next attempt.

Question 70: E

This question requires you to calculate the adjusted score for Ben for each segment. If Ben has a 50% chance of hitting the segment he is aiming for, we can assume he hits each adjacent segment 25% of the time. Thus:

$$Adjusted\,Score = \frac{Segment\,aimed\,at}{2} + \frac{First\,Adjacent\,Segment}{4} + \frac{Second\,Adjacent\,Segment}{4}$$

$$Adjusted\,Score = \frac{Segment\,aimed\,at}{2} + \frac{Sum\,of\,Adjacent\,Segments}{4}$$

E.g. if he aims at segment 1: He will score $\frac{1}{2} + \frac{18+20}{4} = 10$

Now it is a simple case of trying the given options to see which segment gives the highest score. In this case, it is segment 19: $\frac{19}{2} + \frac{7+3}{4} = 12$

Question 71: C

The total cost is £8.75, and Victoria uses a £5.00 note, leaving a total cost of £3.65 to be paid using change.
Up to 20p can be paid using 1p and 2p pieces, so she could use 20 1p coins to make up this amount.
Up to 50p can be paid using 5p and 10p pieces, so she could use 10 5p pieces to make up this amount. This gives a total of 30 coins, and a total payment of £0.70.
Up to £1.00 can be paid using 20p pieces and 50p pieces. Thus, she could use up to 5 20p pieces, giving a total of 35 coins used, and a total payment of £1.70.

The smallest denomination of coin that can now be used is a £1.00 coin. Hence Victoria must use 2 £1.00 coins, giving a total of 37 coins, and a total payment of £3.70. However, we know that the total cost to pay in change was £3.65, and that Victoria paid the exact amount, receiving no change. Thus, we must take away coins to the value of 5p, removing the smallest number of coins possible. This is achieved by taking away 1 5p piece, giving a grand total of 36 coins.

Question 72: B

The time could be 21:25, if first 2 digits were reversed by the glass of water (21 would be reversed to give 15). **A** cannot be the answer, because this would involve altering the last 2 digits, and we can see that 25 on a digital clock, when reversed simply gives 25 (the 2 on the left becomes a 5 on the right, and the 5 on the right becomes a 2 on the left). **C** cannot be the answer, as this involves reversing the middle 2 digits. As with the right two digits, the middle 2 digits of 2:5 would simply reverse to give itself, 2:5. **D** could be the time if the *2nd* and *4th* digits were reversed, as they would both become 2's. However, the question says that 2 *adjacent* digits are reversed, meaning that the 2nd and 4th digits cannot be reversed as required here. **E** is not possible as it would require all four numbers to be reversed.

Thus, the answer is **B**.

Question 73: B

We can see from the question that Lorkdon is a democracy and therefore cannot have been invaded by a democracy because of the treaty (we are assuming this treaty is upheld, as said in the question). Thus, Nordic (which has invaded Lorkdon) *must* be a dictatorship. Now, we can see that Worsid has been invaded by a dictatorship, *and* has invaded a dictatorship. The question states that no dictatorship has undergone both of these events. Thus, we know that Worsid cannot be a dictatorship. We also know from the question that each of these countries is *either* a dictatorship or a democracy. Thus, Worsid must be a democracy.

Question 74: C

The total price of all of these items would usually be £17. However, with the DVD offer, the customer saves £1, giving a total cost of £16. Thus, the customer will need to receive £34 in change.

Question 75: B

To answer this, we simply calculate how much total room in the pan will be taken up by the food for each guest:

- 2 rashers of bacon, giving a total of 14% of the available space.
- 4 sausages, taking up a total of 12% of the available space.
- 1 egg takes up 12% of the available space.

Adding these figures together, we see that each guest's food takes up a total of 38% of the available space.

Thus, Ryan can only cook for 2 guests at once, since 38% multiplied by 3 is 114%, and we cannot use up more than 100% of the available space in the pan.

Question 76: C

To calculate this, let the total number of employees be termed "Y".

We can see that £60 is the total cost for providing cakes for 40% of "Y".

We know that £2 is required for each cake. Thus, we can work out that 30 must be 40% of Y.

$0.4Y = 60/2$

$0.4Y = 30$

$Y = 75$

Thus, we can calculate that the total number of employees must be 75.

Question 77: E

The normal waiting time for treatment is 3 weeks. However, the higher demand in Bob's local district mean this waiting time is extended by 50%, giving a total of 4.5 weeks.

Then, we must consider the delay induced because Bob is a lower risk case, which extends the waiting time by another 20%. 20% of 4.5 is 0.9, so there is a delay of another 0.9 weeks for treatment.

Thus, Bob can expect to wait 5.4 weeks for specialist treatment on his tumour.

Question 78: D

In the class of 30, 40% drink alcohol at least once a month, which is 12. Of these, 75% drink alcohol once a week, which is 9. Of these, 1 in 3 smoke marijuana, which is 3.

In the class of 30, 60% drink alcohol less than once a month, which is 18. Of these, 1 in 3 smoke marijuana, which is 6.

Therefore the total number of students who smoke marijuana is 3+6, which is 9.

Question 79: C

The sequence can either be thought of as doubling the previous number then adding 2, or adding 1 then doubling. Double 46 is 92, plus 2 is 94.

Question 80: B
If the mode of 5 numbers of 3, it must feature at least two threes. If the median is 8, we know that the 3rd largest number is an 8. Hence we know that the 3 smallest numbers are 3, 3, and 8. Because the mean is 7, we know that the 5 numbers must add up to 35. The three smallest numbers add up to 14. Hence the two largest must add up to 21.

Question 81: E
The biggest difference in the weight of potatoes will be if the bag with only 5 potatoes in weighs the maximum, 1100g, and the bag with 10 potatoes weighs the minimum, 900g. If there are 5 equally heavy potatoes in a bag weighing 1100g, each weighs 220g. If there are 10 equally heavy potatoes in a 900g bag, each weighs 90g. The difference between these is 130g.

Question 82: D
There are 60 teams, and 4 teams in each group, so there are 15 groups. In each group, if each team plays each other once, there will be 6 matches in each group, making a total of 90 matches in the group stage. There are then 16 teams in the knockout stages, so 8 matches in the first round knockout, then 4, then 2, then 1 final match when only two teams are left. Hence there are 105 matches altogether (90 + 8 + 4 + 2 + 1 = 105).

Question 83: A
We know the husband's PIN number must be divisible by 8 because it has been multiplied by 2 3 times and had a multiple of 8 added to it. The largest 4 digit number which is divisible by 8 is 9992. Minus 200 is 9792. Divide by 2 is 4896. Hence the largest the husband's last 4 card digits can be is 4896. Minus 200 is 4696. Divide by 2 is 2348. Hence the largest my last 4 card digits can be is 2348. Minus 200 is 2148. Divide by 2 is 1074. Hence the largest my PIN number can be is 1074.

Question 84: C
If the first invitation is sent as early as possible, it will be sent on the 50th birthday. It will be accepted after 2 reminders and hence conducted at 50 years 11 months. The time between each screening will be 3 years 11 months. Hence, the second screening will be at 54 years 10 months. The third screening will be at 58 years 9 months. Hence, the fourth screening will be at 62 years 8 months.

Question 85: A
Ellie has worked for the company for more than five but less than six whole years. At the end of each whole year she receives a pay rise in thousands equal to the number of years of her tenure. Therefore at the end of the first year the raise is £1,000, then at the end of the second year it is £2,000 and so on to year 5. Thus the total amount of her pay comprised by the pay rises is £15,000, so the basic pay before accounting for these rises was £40,000 - £15,000 = £25,000.

Question 86: B
The trains come into the station together every 40 minutes, as the lowest common multiple of 2, 5 and 8 is 40. Hence, if the last time trains came together was 15 minutes ago, the next time will be in 25 minutes.

Question 87: C
If you smoke, your risk of getting Disease X is 1 in 24. If you drink alcohol, your risk of getting Disease X is 1 in 6. Each tablet of the drug halves your risk. Therefore a drinker taking 1 tablet means their risk is 1 in 12, and taking 2 tablets means their risk is 1 in 24, the same as someone who smokes.

Question 88: A
There are 10 red and 8 green balls. Clearly the most likely combination involves these colours only. Since there are more red balls than green, the probability of red-red is greater than green-green. However, there are **two** possible ways to draw a combination, either the red first followed by green or green first followed by red. The probability of red-red = $\left(\frac{10}{20} x \frac{9}{19}\right) = \frac{9}{38}$.

The probability of red and green = $\left(\frac{8}{20} x \frac{10}{19}\right) + \left(\frac{10}{20} x \frac{8}{19}\right) = \frac{8}{38} + \frac{8}{38} = \frac{16}{38}$. Therefore the combination of red and green is more likely.

Question 89: B
The least likely combination of balls to draw is blue and yellow. You are much more likely to draw a green ball than either a blue or yellow one because there are many more in the bag. Since the draw is taken without replacement, yellow and yellow is impossible because there is only one yellow ball.

Question 90: F
Since there is only 1 blue and 1 yellow ball, it is possible to take 18 balls which are red or green. You would need to take 19 of the 20 balls to be certain of getting either the blue ball or the yellow ball.

Question 91: C
The smallest number of parties required would theoretically be 3 – Namely Labour, the Liberal Democrats and UKIP, giving a total of 355 seats. However, the Liberal Democrats will not form a coalition with UKIP, so this will not be possible. Thus, there are 2 options:

- Labour can form a coalition with the Greens and UKIP, which is not contradictory to anything mentioned in the question. This would give a total of 325 seats, and would thus need the next 2 largest parties (The Scottish National Party and Plaid Cymru) in order to get more than 350 seats, meaning 5 parties would need to be involved.
- Alternatively, Labour can form a coalition with the Liberal Democrats and the Green Party. This would give a total of 340 seats. Only one more party (e.g. the Scottish National Party) would be required to exceed 350 seats, giving a grand total of 4 parties.

Thus, the smallest number of parties needed to form a coalition would be 4.

Question 92: E
360 appointments are attended and only 90% of those booked are attended, meaning there were originally 400 appointments booked in and 40 have been missed. 1 in 2 of the booked appointments were for male patients, so 200 appointments were for male patients. Male patients are three times as likely to miss booked appointments, so of the 40 that were missed, 30 were missed by men. Given that of 200 booked appointments, 30 were missed, this means 170 were attended.

Question 93: B
If every one of 60 students studies 3 subjects, this is 180 subject choices altogether. 60 of these are Maths, because everyone takes Maths. 60% of 60 is 36, so 36 are Biology. 50% of 60 is 30, so 30 are Economics and 30 are Chemistry. 60+36+30+30=156, so there are 24 subject choices left which must be Physics.

Question 94: B
If 100,000 people are diagnosed with chlamydia and 0.6 partners are informed each, this is 60,000 people, of which 80% (so 48,000) have tests. 12,000 of the partners who are informed, as well as 240,000 who are not (300,000 – 60,000) do not have tests. This makes 252,000 who are not tested. We can assume that half of these people would have tested positive for chlamydia, which is 126,000. So the answer is 126,000.

Question 95: C
Tiles can be added at either end of the 3 lines of 2 tiles horizontally or at either end of the 2 lines of 2 tiles vertically. This is a total of 10, but in two cases these positions are the same (at the bottom of the left hand vertical line and the top of the right hand vertical line). So the answer is 10 – 2 = 8.

Question 96: C
Harry needs a total of 4000ml + 1200ml = 5200ml of squash. He has 1040ml of concentrated squash, which is a fifth of the total dilute squash he needs. So he will need 4 parts water to every 1 part concentrated squash, therefore the resulting liquid is 1/5 squash and 4/5 water.

Question 97: C
There are 24 different possible arrangements (4 x 3 x 2 x 1), which means that there are 23 other possible arrangements than Alex, Beth, Cathy, Daniel.

Question 98: F
A is incorrect because the distance travelled is only 10 miles. **B** is incorrect because the distance travelled is 19 miles. **C** is incorrect because no town is visited twice. **D** is incorrect because Hondale and Baleford are both visited twice. **E** is incorrect because no town is visited twice. Therefore **F** is the correct answer.

Question 99: C
Georgia is shorter than her Mum and Dad, and each of her siblings is at least as tall as Mum (and we know Mum is shorter than Dad because Ellie is between the two), so we know Georgia is the shortest. We know that Ellie, Tom and Dad are all taller than Mum, so Mum is second shortest. Ellie is shorter than Dad and Tom is taller than Dad, so we can work out that Ellie must be third shortest.

Question 100: A
Danielle must be sat next to Caitlin. Bella must be sat next to the teaching assistant. Hence these two pairs must sit in different rows. One pair must be sat at the front with Ashley, and the other must be sat at the back with Emily. Since the teaching assistant has to sit on the left, this must mean that Bella is sat in the middle seat and either Ashley or Emily (depending on which row they are in) is sat in the right hand seat. However, Bella cannot sit next to Emily, so this means Bella and the teaching assistant must be in the front row. So Ashley must be sat in the front right seat.

Question 101: C
The dishwasher is run 2+p times a week, where *p* is the number of people in the house. Let the number of people in the house when the son is not home be *s*, and when the son is home it is s+1. In 30 weeks when the son is home, she would buy 6 packs of dishwasher tablets. In 30 weeks when the son is not home, she would buy 5 packs of dishwasher tablets. So 1.2 times as many packs of dishwasher tablets are bought when he is home. So 2+s+1 is 1.2 time 2+s.
i.e. $2.4 + 1.2s = 2 + s + 1$
Therefore 0.2s = 0.6
s = 3
When her son is home, there are s + 1 = 4 people in the house.

Question 102: A
No remaining days in the year obey the rule. The next date that does is 01/01/2015 (integers are 0, 1, 2, 5). This is 6 days later than the specified date.

Question 103: B
If each town is due North, South, East or West of at least 2 other towns and we know that one is east and one is north of a third, then they must be arranged in a square. So Yellowtown is 4 miles east of Bluetown to make a square, which means it must be 5 miles north of Redtown. So Redtown is 5 miles south of Yellowtown.

Question 104: B
Jenna pours 4/5 of 250 ml into each glass, which is 200 ml. Since she has 1500 ml of wine, she pours 100 ml into the last glass, which is 2/5 of the 250 ml full capacity.

Question 105: E
The maximum number of girls in Miss Ellis's class with brown eyes and brown hair is 10, because the two thirds of the girls with brown eyes could also all have brown hair. The minimum number is 0 because it could be that all the boys, and the third of the girls without brown eyes, all had brown hair, which would be 2/3 of the class.

Question 106: E
A negative "score" results from any combination of throws which includes a 1 but from no other combination. Given that a negative score has a 0.75 probability, a positive or zero score has a 0.25 probability. Therefore throwing two numbers that are not 1 twice in a row has a probability of 0.25. Hence, the probability of throwing a non-1 number on each throw is $\sqrt{0.25} = 0.5$. So the probability of throwing a 1 on an individual throw is $1 - 0.5 = 0.5$.

Question 107: C

We can work out from the information given the adult flat rate and the charge per stop. Let the charge per stop be s and the flat rate be f. Therefore: 15s + f = 1.70

8s + f = 1.14

We can hence work out that: 7s = 0.56, so s = 0.08. Hence, f = 0.50

Megan is an adult so she pays this rate. For 30 stops, the rate will be 0.08 x 30 + 0.50 = 2.90.

Question 108: B

We found in the previous question that the flat rate for adults is £0.50 and the rate per stop is £0.08. We know that the child rate is half the flat rate and a quarter of the "per stop" rate, so the child flat rate is £0.25 and the rate per stop is 2p. So for 25 stops, Alice pays:

0.02 x 25 + 0.25 = 0.75

Question 109: C

We should first work out how many stops James can travel. For £2, he can afford to travel as many stops as £1.50 will take him once the flat rate is taken into account. The per stop rate is 8p per stop, so he can travel 18 stops, so he will need to go to the 18th stop from town. So he will need to walk past 7 stops to get to the stop he can afford to travel from.

Question 110: D

The picture will need a 12 inch by 16 inch mount, which will cost £8. It will need a 13 inch by 17 inch frame, which will cost £26. So the cost of mounting and framing the picture will be £8 + £26 = £34.

Question 111: C

Mounting and framing an 8 by 8 inch painting will cost £5 for the mount and £22 for the frame, which is £27. Mounting and framing a 10 by 10 inch painting will cost £6 for the mount and £26 for the frame, which is £32. The difference is £32 - £27 = £5.

Question 112: B

We found in the last question that mounting and framing a 10 by 10 inch painting will cost £6 for the mount and £26 for the frame, which is £32 total. We can calculate that each additional inch of mount and frame for a square painting costs £2.50; £2 for the frame and £0.50 for the mount. So an 11 inch painting will cost £34.50 to frame and mount, a 12 inch £37, a 13 inch £39.50, a 14 inch £42. The biggest painting that can be mounted and framed for £40 is a 13 inch painting.

Question 113: D

Recognise that the pattern is *"consonants move forward by two consonants; vowel stay the same"*. This allows coding of the word MAGICAL to PAJIFAN to RALIHAQ.

Forward two			Forward two
M	⇒	O (skips to) P	⇒ R
A	⇒	Stays the same	⇒ A
G	⇒	I (skips to) J	⇒ L
I	⇒	Stays the same	⇒ I
C	⇒	E (skips to) F	⇒ H
A	⇒	Stays the same	⇒ A
L	⇒	N	⇒ Q

Question 114: C

If *f* donates the flat rate, and *k* denotes the rate per km, we can form simultaneous equations:

f + 5k = £6 AND f + 3k = £4.20

Subtract equation two from equation one:

(f + 5k) - (f + 3k) = £6 - £4.20

Thus, 2k = £1.80 and k = £0.90

Therefore, f + (5 x 0.90) = £6

So, f + £4.50 = £6. Thus, f = £1.50

7k will be £1.50 + 7 x £0.90 = £7.80

Question 115: C

The increase from 2001/2 to 2011/12 was 1,019 to 11,736, which equals a linear increase of 10,717 admissions.

So, in 20 years, we would expect to see an increase by 10,717 x 2 = 21,434. Add this to the number in 2011 to give 33,170 admissions.

Question 116: A

As the question uses percentages, it does not matter what figure you use. To make calculations easier, use an initial price of £100. When on sale, the dress is 20% off, so using a normal price of £100, the dress would be £80. When the dresses are 20% off, the shop is making a 25% profit. Therefore: £80 = 1.25 x purchase price.

Therefore, the purchase price is: $\frac{80}{1.25}$ = £64. Thus, the normal profit is £100 - £64 = £36. I.e. when a dress sells for £100, the shop makes £36 or 36% profit.

Question 117: C

1. Incorrect. There must be 6 general committee clinical students, plus the treasurer, and 2 sabbatical roles, none of whom can be preclinical, so there must be a maximum of 11 preclinical students.
2. Correct. There must two general for each year plus welfare and social officers, totalling to 6.
3. Incorrect. The committee is made up of 20 students, 2 roles are sabbatical, so there are 18 studying students, and therefore there can be 3 from each year.
4. Correct. There are 18 studying students on the committee, and there must be 6 general committee members from pre-clinical, plus welfare and social, therefore there must be a minimum of 8 pre-clinical students, so there must be 10 clinical students.
5. Incorrect. You need to count up the number of specific roles on the committee, which is 5, and there must be 2 students from each year, which is 12. This leaves 3 more positions, which the question doesn't state can't be first years. Therefore there could be up to 5 first years.
6. Incorrect. There must be at least 2 general committee members from each year. However, the worked answer to 5 shows there are 15 general committee members which are split across the 6 years, and so there must be an uneven distribution.

Question 118: B

Remember 2012 was a leap year. Work through each month, adding the correct number of days, to work out what day each 13th would be on.

If a month was 28 days, the 13th would be the same day each month, therefore to work this out quickly, you only need to count on the number of days over 28. For example, in a month with 31 days, the 13th will be 3 weekdays (31-28) later.

Thus if 13th January is a Friday, 13th February is a Monday, (February has 29 days in 2012), 13th March is a Tuesday and 13th April is a Friday.

Question 119: E

There are 18 sheep in total. The question states there are 8 male sheep, which means there are 10 female sheep before some die. 5 female sheep die, so there are 5 female sheep alive to give birth to lambs. Each delivers 2 lambs, making 10 lambs in total. There are 4 male sheep and 5 mothers so the total is 10 + 4 + 5 = 19 sheep.

Question 120: D

We can see from the fact that all the possible answers end "AME" that the letters "AME" must be translated to the last 3 letters of the coded word, "JVN", under the code. J is the 10th letter of the alphabet so it is 9 letters on from A (V is the 21st letter of the alphabet and M is the 13th, and N is the 14th letter of the alphabet and E is the 5th, therefore these pairs are also 9 letters apart). Therefore P is the code for the letter 9 letters before it in the alphabet. P is the 16th letter of the alphabet, therefore it is the code for the 7th letter of the alphabet, G. Therefore from these solutions the only possibility for the original word is GAME.

Question 121: C

Let x be the number of people who get on the bus at the station.

It is easiest to work backwards. After the 4th stop, there are 5 people on the bus. At the 4th stop, half the people who were on the bus got off (and therefore half stayed on) and 2 people got on. Therefore, 5 is equal to 2 plus half the number of people who were on the bus after the 3rd stop. So half the number of people who were on the bus after the 3rd stop must be 3. Therefore, after the 3rd stop, there must have been 6 people on the bus.

We can then say that 6 is equal to 2 plus half the number of people who were on the bus after the 2nd stop. Therefore there were 8 people on the bus after the 2nd stop.

We can then say that 8 is equal to 2 plus half the number of people who were on the bus after the 1st stop. Therefore there were 12 people on the bus after the 1st stop.

We can then say that 12 is equal to 2 plus half the number of people who got on the bus at the station. Therefore the number of people who got on the bus at the station is 20.

Question 122: B

We know from the question that I have purchased small cans of blue and white paint, and that blue paint accounted for 50% of the total cost. Since a can of blue paint is 4 X the price of a can of white paint, we know I must have purchased 4 cans of white paint for each can of blue paint.

Each can of small paint covers a total of $10m^2$, and I have painted a total of $100m^2$, in doing so using up all the paint. Therefore, I must have purchased 10 cans of paint. Therefore, I must have purchased 2 cans of blue paint and 8 cans of white paint. So I must have painted $20m^2$ of wall space blue.

Question 123: E

The cost for x cakes under this offer can be expressed as: $x(42-x^2)$

Following this formula, we can see that 2 cakes would cost 76p, 3 cakes would cost 99p, and 4 cakes would cost 104p. As the number of cakes increases beyond 4, we see that the overall price actually drops, as 5 cakes would cost 85p and 6 cakes would cost 36p. This confirms Isobel's prediction that the offer is a bad deal for the baker, as it ends up cheaper for the customer to purchase more cakes. It is clear that 6 cakes is the smallest number for which the price will be under 40p, and the price will continue to drop as more cakes are purchased.

Question 124: C

Adding up the percentages of students in University A who do "Science" subjects gives:

$23.50 + 6.25 + 30.25 = 60\%$.

60% of 800 students is 480, so 480 students in University A do "Science" subjects.

Adding up the percentages of students in University B who do "Science subjects" gives:

$13.25 + 14.75 + 7.00 = 35\%$. 35% of 1200 students is 420, so 420 students in University B do "Science" subjects. Therefore:

$480 - 420 = 60$

60 more students in University A than University B take a "Science" subject.

Question 125: C

Let the number of miles Sonia is travelling be x. Because she is crossing 1 international border, travelling by Traveleasy Coaches will cost Sonia: £(5 + 0.5x)

Travelling by Europremier coaches will cost Sonia: £(15 + 0.1x).

Because we know the cost is the same for both companies, the number of miles she is travelling can be found by setting these two expressions equal to each other: $5 + 0.5x = 15 + 0.1x$.

This equation can be rearranged to give: $0.4x = 10$

Therefore: $x = 10/0.4 = 25$

Question 126: E
To find out whether many of these statements are true it is necessary to work out the departure and arrival times, and journey time, for each girl.

Lauren departs at 2:30pm and arrives at 4pm, therefore her journey takes 1.5 hours
Chloe departs at 1:30pm and her journey takes 1 hour longer than 1.5 hours (Lauren's journey), therefore her journey takes 2.5 hours and she arrives at 4pm
Amy arrives at 4:15pm and her journey takes 2 times 1.5 hours (Lauren's journey), therefore her journey takes 3 hours and she departs at 1:15pm.
Looking at each statement, the only one which is definitely true is **E**: Amy departs at 1:15pm and Chloe departs at 1:30pm therefore Amy departed before Chloe.

D *may* be true, but nothing in the question shows it is *definitely* true, so it can be safely ignored.

Question 127: B
First consider how many items of clothing she can take by weight. The weight allowance is 20kg. Take off 2kg for the weight of the empty suitcase, then take off another 3kg (3 X 1000g) for the books she wishes to take. Therefore she can fit 15kg of clothes in her suitcase. To find out how many items of clothing this is, we can divide 15kg=15000g by 400g: 15000/400 = 150/4 = 37.5
So she can pack up to 37 items of clothing by weight.

Now consider the volume of clothes she can fit in. The total volume of the suitcase is:
50cm x 50cm x 20cm = 50000cm^3
The volume of each book is: 0.2m x 0.1m x 0.05m = 1000cm^3

So the volume of space available for clothes is: 50000 – (3 x 1000) = 47000cm^3
To find out how many items of clothing she can fit in this space, we can divide 47000 by 1500:
47000/1500 = 470/15 = 31 1/3
So she can pack up to 31 items of clothing by volume.

Although she can fit 37 items by weight, they will not fit in the volume of the suitcase, so the maximum number of items of clothing she can pack is 31.

Question 128: D
We can work out the Answer by considering each option:
Bed Shop A: £120 + £70 = £190
Bed Shop B: £90 + £90 = £180
Bed Shop C: £140 + (1/2 x £60) = £170
Bed Shop D: (2/3) x (£140+£100) = (2/3) x (£240) = £160
Bed Shop E: £175
Therefore the cheapest is Bed Shop **D**.

Question 129: C
The numbers of socks of each colour is irrelevant, so long as there is more than one of each (which there is). There are only 4 colours of socks, so if Joseph takes 5 socks, it is guaranteed that at least 2 of them will be the same colour.

Question 130: D
Paper comes in packs of 500, and with each pack 20 magazines can be printed. Each pack costs £3.
Card comes in packs of 60, and with each pack 60 magazines can be printed. Each pack costs £3 x 2 = £6.
Each ink cartridge prints 130 sheets, which is 130/26 = 5 magazines. Each cartridge costs £5.

The lowest common multiple of 20, 60 and 5 is 60, so it is possible to work out the total cost for printing 60 magazines. Printing 60 magazines will require 3 packs of paper at £3, 1 pack of card at £6 and 12 ink cartridges at £5. So the total cost of printing 60 magazines is: (3 x 3) + 6 + (12 x 5) = £75.

The total budget is £300.
£300/£75 = 4
So we can print 4x60 magazines in this budget, which is 240 magazines.

Question 131: E
We can express the information we have as: $\frac{1}{4} - \frac{1}{5} = \frac{1}{20}$
So the six additional lengths make up 1/20 of Rebecca's intended distance. So the number of lengths she intended to complete was: 20 x 6 = 120.

Question 132: B
Sammy has a choice of 3 flavours for the first sweet that he eats. Each of the other sweets he eats cannot be the same flavour as the sweet he has just eaten. So he has a choice of 2 flavours for each of these four sweets. So the total number of ways that he can make his choices is:
$3 \times 2 \times 2 \times 2 \times 2 = 48$

Question 133: C
Suppose that today Gill is x years old. It follows that Granny is 15x years old. In 4 years' time, Gill will be (x+4) years old and Granny will be 15x+4 years old. We know that in 4 years' time, Granny's age is equal to Gill's age squared, so: $15x + 4 = (x + 4)^2$
Expanding and rearranging, we get: $x^2 - 7x + 12 = 0$
We can factorise this to get: $(x - 3)(x - 4)$
So x is either 3 or 4. Gill's age today is either 3 or 4 so Granny is either 45 or 60. We know Granny's age is an even number, so she must be 60 and hence Gill must be 4. So the difference in their ages is 56 years.

Question 134: C
If Pierre is telling the truth, everyone else is not telling the truth. But, also in this case, what Qadr said is not true, and hence Ratna is telling the truth. So we have a contradiction. So we deduce that Pierre is not telling the truth. Therefore, Qadr is telling the truth, and so Ratna is not telling the truth. So Sven is also telling the truth, and hence Tanya is not telling the truth. So Qadr and Sven are telling the truth and the other three are not telling the truth.

Question 135: D
Angus walks for 20 minutes at 3 mph and runs for 20 minutes at 6 mph. 20 minutes is one-third of an hour. So the number of miles that Angus covers is: $3 \times 13 + 6 \times 13 = 6$
Bruce covers the same distance. So Bruce walks 12×3 miles at 3 mph which takes him 30 minutes and runs the same distance at 6 mph which takes him 15 minutes. So altogether it takes Bruce 45 minutes to finish the course.

Question 136: B
Although you could do this quickly by forming simultaneous equations, it is even quicker to note that 72 x 4 = 288. Since Species 24601 each have 4 legs; it leaves a single member of species 8472 to account for the other 2 legs.

Question 137: E
None of the options can be concluded for certain. We are not told whether any chicken dishes are spicy, only that they are all creamy. Whilst all vegetable dishes are spicy, some non-vegetable dishes could also be spicy. There is no information on whether dishes can be both creamy and spicy, nor on which, if any, dishes contain tomatoes. Remember, if you're really stuck, draw a Venn diagram for these types of questions.

Question 138: C
At 10mph, we can express the time it takes Lucy to get home as: 60 x 8/10 = 48
Since Simon sets off 20 minutes later, his time taken to get home, in order to arrive at the same time, must be: 48 – 20 = 28
Therefore his cycling speed must be: 48/28 x 10 = 17mph

Question 139: A
The total profit from the first transaction can be expressed as: 2000 x 8 = 16,000p
The total profit from the second transaction is: 1000 x 6 = 6,000p

Therefore the total profit is 22,000p or £220 before charges. There are four transactions at a cost of £20 each, therefore the overall profit is: £220 – (20 x 4) = £140

Question 140: C
For the total score to be odd, there must be either three odd or one odd and two even scores obtained. Since the solitary odd score could be either the first, second or third throw there are four possible outcomes that result in an odd total score. Additionally, there are the same number of possibilities giving an even score (either all three even or two odd and one even scores obtained), and the chance of throwing odd or even with any given dart is equal. Therefore, there is an equal probability of three darts totalling to an odd score as to an even score, and so the chance of an odd score is ½.

Question 141 C
This is a compound interest question. £5,000 must be increased by 5%, and then the answer needs to be increased by 5% for four more iterations. After one year: £5,000 x 1.05 = £5,250
Increasing sequentially gives 5512, 5788, 6077 and 6381 after five years. Therefore the answer is £6,381.

Question 142: D
If in 5 years' time the sum of their ages is 62, the sum of their ages today will be: 62 – (5 x 2) = 52
Therefore if they were the same age they would both be 26, but with a 12 year age gap they are 20 and 32 today. Michael is the older brother, so 2 years ago he would have been aged 30.

Question 143: A
Tearing out every page which is a multiple of 3 removes 166 pages. All multiples of 6 are multiples of 3, so no more pages are torn out with that instruction. Finally, half of the remaining pages are removed, which equates to an additional 167 pages. Therefore 333 pages are removed in total. The total surface area of these pages is 15 x 30 x 333 = 149,850 cm^2 = 14.9m^2. At 110 gm^2, 14.9 m^2 weighs 14.9 x 110 = 1,650g (1,648g unrounded)

Question 144: D
The cost of fertiliser is 80p/kg = 8p/100g. At 200g the incremental increase in yield is 65 pence/m. At each additional 100g it will be reduced by 30%, therefore at 300g/m it is 45.5p, at 400g/m it is 31.8p, at 500g/m it is 22.3p, at 600g/m it is 15.6p, at 700g/m it is 10.9p, and at 800g it is 7.6p. So at 800g the gain in yield is less than the cost of the fertiliser to produce the gain, and so it is no longer cost effective to fertilise more.

Question 145: D
Statements **A**, **C** and **E** are all definitely true. Meanwhile, statement **B** may be not true but is not definitely untrue, as this depends on the number of cats and rabbit owned.
Only statement **D** is definitely untrue. The type of animal requiring the most food is a dog, and as can be seen from the tables, Furry Friends actually sells the most expensive dog food, not the cheapest.

Question 146: C
The largest decrease in bank balance occurs between January 1st and February 1st, totalling £171, reflecting the amount spent during the month of January, £1171. However, because there is a pay rise beginning on March 10th, we need to consider that from April onwards, the bank balance will have increased by £1100, not £1000. This means that the same decrease in bank balance reflects £100 more spending if it occurs after March. This means that 2 months now have seen more spending than February. Between March 1st and April 1st, the bank balance has decreased by £139. With the salary increase, the salary is now £1100, so the total spending for the month of March is £1239. This is greater than the total spending during the month of January.
Similarly, the month of April has also seen more spending than January once the pay rise is considered, a total of £1225 of spending. However, this is still less than the month of March.

Question 147: C
If Amy gets a taxi, she can set off 100 minutes before 1700, which is 1520.
If Amy gets a train, she must get the 1500 train as the later train arrives after 1700, so she must set off at 1500.

Since Northtown airport is 30 minutes from Northtown station, there is no way Amy can get the flight and still arrive at Northtown station by 1700. Therefore Amy should get a taxi and should leave at 1520.

Question 148: C
We can decompose the elements of the multiplication grid into their prime factors, thus:

	C	**D**
A	2 x 2 x 2 x 3 x 7	2 x 2 x 2 x 2 x 3 x 3 x 5
B	7 x 17	2 x 3 x 5 x 17

bc = 7 x 17, so one of b and c must be 7 and the other must be 17. b must be 17 because bd is a multiple of 17 and not of 7, and c must be 7 because ac is a multiple of 7 and not of 17. ac is 168, so a must be 168 divided by 7, which is 24. ad is 720 so d must be 720 divided by 24, which is 30. Hence the answer is 30.
Alternatively approach the question by eliminating all answers which are not factors of both 720 and 510.

Question 149: E
48% of the students are girls, which is 720 students. Hence 80 is 1/9 of the girls, so 1/9 of boys are mixed race. The remaining 780 students are boys, so 87 boys are mixed race to the nearest person. There is a shortcut to this question. Notice that 80 girls are mixed race, and the proportion is the same for boys. As there are more boys than girls we know the answer is greater than 80. Option **E** 90 is the only option for which this holds true.

Question 150: D
Don't be fooled – this is surprisingly easy. We can see that between Monday and Thursday, Christine has worked a total of 30 hours. We can also calculate how long her shift on Friday was supposed to be. She is able to make up the hours by working 3 extra hours next week, and 5 hours on Sunday. Thus, the Friday shift must have been planned to be 8 hours long. Adding this to the other 30 hours, we see that Christine was supposed to work 38 hours this week.

Question 151: C
130°. Each hour is 1/12 of a complete turn, equalling 30°. The smaller angle between 8 and 12 on the clock face is 4 gaps, therefore 120°. In addition, there is 1/3 of the distance between 3 and 4 still to turn, so an additional 10° must be added on to account for that.

Question 152: B
The total price of all of these items would usually be £17. However, with the DVD offer, the customer saves £1, giving a total cost of £16. Thus, the customer will need to receive £34 in change.

Question 153: E
A. Incorrect. UCL study found eating more portions of fruit and vegetables was beneficial.
B. Incorrect. This is a possible reason but has yet to be fully investigated.
C. Incorrect. Fruit and vegetables are more protective against cardiovascular disease, and were shown to have little effect on cancer rates.
D. Incorrect. Inconclusive – people who ate more vegetables generally had a lower mortality but unknown if this is due to eating more vegetables or other associated factors.
E. Correct. Although this has previously been the case, this study did not find so. 'they recorded no additional decline for people who ate over 5 portions'.
F. Incorrect. The 5% decline per portion was only up to 5 portions and no additional reduction in mortality for 7 than 5 portions.
G. Incorrect. Study only looks at cancers in general and states need to look into specific cancers.

Question 154: C

Deaths in meta-analysis = 56423/800000 = 0.07 or 7%
1% lower in UCL study so 6%
6% of 65,000 = 65000 x 0.06 = 3,900

Question 155: B

A. Eating more fruit and vegetables doesn't particularly lower overall risk but need research into specific cancer risk.
B. The UCL research alone found that increasing the number of fruit and vegetable portions had a beneficial effect, even though this wasn't the overall conclusion when combined with results from the meta-analysis.
C. The results were not exactly the same but showed similar overall trends.
D. Although this may be true, there is no mention of this in the passage.
E. Fruit and vegetables are protective against cardiovascular disease, but not exclusively. They also reduce the rates of death from all causes.
F. The UCL study is in England only and the meta-analysis a combination of studies from around the world.
G. Suggested by the UCL research, but not the meta-analysis, so not an overall conclusion of the article.

Question 156: E

Remember that you don't need to calculate exact values for question 249 – 251. Thus, you should round numbers frequently to make this more manageable. Work out percentage of beer and wine consumption and then the actual value using the total alcohol consumption figure:

Belarus: 17.3 + 5.2 = 22.5%;
0.225 x 17.5 = 3.94

Lithuania: Missing figure 100 – 7.8 – 34.1 – 11.6 = 46.5
46.5 + 7.8 = 54.3%
0.543 x 15.4 = 8.36

France: 18.8 + 56.4 = 75.2%
0.752 x 12.2 = 9.17

Ireland: 48.1 + 26.1 = 74.2
0.742 x 11.9 = 8.83

Andorra: missing figure 100 – 34.6 – 20.1 = 45.3
34.6 + 45.3 = 79.9%
0.799 x 13.8 = 11.0

Question 157: D

Russia:
2010 – Total = 11.5+3.6 = 15.1. Spirits = 0.51 x 15.1 = 7.7
2020 – Total = 14.5. Spirits = 0.51 x 14.5 = 7.4
Difference = 0.3 L

Belarus:
2010 – Total = 14.4 + 3.2 = 17.6. Spirits = 0.466 x 17.6 = 8.2
2020 – Total = 17.1. Spirits = 0.466 x 17.1 = 8.0
Difference = 0.2 L

Lithuania:
2010 – Total = 15.4. Spirits = 0.341 x 15.4 = 5.3
2020 – Total = 16.2. Spirits = 0.341 x 16.2 = 5.5
Difference = 0.2 L

Grenada:
2010 – Total = 12.5. Spirits % = 100 – 29.3 – 4.3 – 0.2 = 66.2%. Spirits = 0.662 x 12.5 = 8.3
2020 – Total = 10.4. Spirits = 0.662 x 10.4 = 6.8
Difference = 1.5 L

Ireland:
2010 – Total = 11.9. Spirits = 0.187 x 11.9 = 2.2
2020 – Total = 10.9. Spirits = 0.187 x 10.9 = 2
Difference = 0.2 L

Question 158: C
Work out 4.9 as a percentage of total beer consumption in Czech Republic and search other rows for similar percentage.

4.9/13 = 0.38, approx. 38% which is very similar to percentage consumption in Russia (37.6).

Question 189: B
We can add up the total incidence of the 6 cancers in men, which is 94,000. Then we can add up the total incidence in women, which is 101,000. As a percentage of 10 million, this is 0.94% of men and 1.01% of women. Therefore the difference is 0.07%.

Question 160: C
Given there are 1.15 times as many men as women, the incidence of each cancer amongst men needs to be greater than 1.15 times the incidence amongst women in order for a man to be more likely to develop it. The incidence is at least 1.15 higher in men for 3 cancers (prostate, lung and bladder).

Question 161: D
If 10% of cancer patients are in Sydney, there are 10,300 prostate/bladder/breast cancer patients and 9,200 lung/bowel/uterus cancer patients in Sydney. Hence the total number of hospital visits is 10,300 + 18,400, which is 28,700.

Question 162: A
The proportion of men with bladder cancer is 2/3 and women 1/3.

Question 163: D
First we work out the size of each standard drink. 50 standard drinks of vodka is equivalent to 1250ml, so one drink is 25ml or 0.025 litres. 11.4 standard drinks of beer is 10 pints of 5700ml, so one standard drink is 500ml or 0.5 litres. 3 standard drinks of cocktail is 750ml so one is 250ml or 0.25 litres. 3.75 standard drinks of wine is 750ml, so one is 200ml or 0.2 litres.

We can then work out the number of units in each drink. Vodka has 0.025 x 40 = 1 unit, Beer has 0.5 x 3 = 1.5 units, Cocktail has 0.25 x 8 = 2 units and Wine has 0.2 x 12.5 = 2.5 units. Since the drink with the most units is wine, the answer is D.

Question 164: B
We found in the last question that vodka has 1 unit, beer has 1.5, cocktail has 2 and wine has 2.5. Hence in the week, Hannah drinks 23.5 units and Mark drinks 29 units. Hence Hannah exceeds the recommended amount by 9.5 units and Mark by 9 units.

Question 165: D
We found that vodka has 1 unit, beer has 1.5, cocktail has 2 and wine has 2.5. Hence it is possible to make 5 combinations of drinks that are 4 units: 4 vodkas, 2 cocktails, 2 vodkas and a cocktail, 1 vodka and 2 beers, or a wine and a beer.

Question 166: D
The total number of males in Greentown is 12,890. Adding up the rest of the age categories, we can see that 10,140 of these are in the older age categories. Hence there are 2750 males under 20.

Question 167: C
Given that in the first question we found the number of males under 20 is 2,750, we can then add up the totals in the age categories (apart from 40-59) in order to find that 15,000 of the residents of Greentown are in other age categories. Hence 9,320 of the population are aged 40-59. We know that 4,130 of these are male, therefore 5,190 must be female.

Question 168: C
The age group with the highest ratio of males:females is 20-39, with approximately 1.9 males per females (approximately 3800:2000). As a ratio of females to males, this is 1:1.9.

Question 169: C
There are 4 instances where the line for Newcastle is flat from one month to the next per year, hence in 2008-2012 (5 years) there are 20 occasions when the average temperature is the same from month to month. During 2007, there are 2 occasions, and during 2013 there are 3.

Question 170: A
The average temperature is lower than the previous month in London for all months from August to December, which is 5 months. However, in August and November in Newcastle, the average temperature remains the same as the previous month. Hence there are only 3 months where the average temperature is lower in both cities. Hence from 2007 to 2012, there are 18 months where the average temperature is lower than the previous month. During 2013, the only included month where the temperature is lower in both cities than the previous month is September. Hence there are 19 months in total when the temperature is lower in both cities than the previous month.

Question 171: B
Firstly work out the difference between average temperatures for each month (2, 3, 1, 2, 1, 3, 3, 2, 2, 5, 1, 0). Then sum them to give 25. Divide by the number of months (12) to give $2^1/_{12}$, which is 2°C to the nearest 0.5°C.

Question 172: D
There is not enough information to tell which month the highest sales are in. We know it increases up to a point and decreases after it, but as we don't know by how much we cannot project where the maximum sales will be.

Question 173: B
Given that by observation, Q2 and Q3 both account for 1/3 of the sales and Q4 accounts for 1/4, this leaves that Q1 accounts for 1/12 of sales. 1/12 of £354,720 is £29,560.

Question 174: A
Quarter 2 accounts for 1/3 of the sales, which is £60,000 in sales revenue. If a tub of ice cream is sold for £2 and costs the manufacturer £1.50, this means profit is 1/4 of sales revenue. Hence £15,000 profit is made during Q2. Hence the answer is A.

Question 175: D

A. and B – Incorrect. Both *could* be true but neither is *definitely* true as it is dependent on the relative number of families with each number of children, which is not given in the question. Therefore we cannot know for certain whether these statements are true.

C – Incorrect. C is definitely *untrue* as half of the families spend £400 a month on food, which totals £4800 a year.

D – Correct. This option is true as 1/6 of families with 1 child and 1/6 of families with 3 children spent £100 a month on food.

E – Incorrect. This option is definitely untrue as the average expenditure for families with 2 children is actually £400 a month.

Question 176: B

2210 out of 2500 filled in responses, meaning that 290 did not. 290 as a percentage of 2500 is roughly 12% (11.6%) of the school that did not respond.

Question 177: C

The percentage of students that saw bullying and reported it was 35%, so 65% of those who saw it did not which is equivalent to 725 students. Of this 725, 146 which roughly equals 20%, gave the reason that they did not think it was important.

Question 178: B

Of the students who told a teacher, 286 did not witness any action. Of those who did notice action, i.e. 110, only 40% noticed any direct action with the bully involved. 40% of 110 is 44, so the correct answer is B.

Question 179: D

"427 cited fears of being found out" which means about 59% out of the 725 students that did not tell about the bullying, cited that it was because they worried about others finding out.

Question 180: F

North-east: 56 per 100,000 on average. This means that there must be a higher propoertion of women than this and a lower proportion of men, such that the average is 56/100,000

We must make the reasonable assumption that there are the same number of men and women in the population as the question asks us to approximate.

Therefore there are 18.6/50,000 men and 37.3/50,000 women

This scales to 74.4/100,000 women which is roughly 74/100,000.

Question 181: C

8 million children – question tells to approximate to 4 million girls and 4 million boys.

Girls: 20% eat 5 portions fruit and vegetables a day. 20% of 4 million: 4 x 0.2 = 0.8 million

Boys: 16% eat 5 portions of fruit and vegetables a day. 16% of 4 million: 4 x 0.16 = 0.64 million

Number of more girls: 800,000 – 640,000 = 160,000.

Question 182: B

A. Incorrect. Women: 13619+10144+6569 = 30332. Men: 16818 + 9726 + 7669 + 6311 = 40524
B. Correct. Flu + pneumonia, lung cancer and chronic lower respiratory diseases = 15361 + 13619 + 14927 = 43907
C. Incorrect. More common cause of death but no information surrounding prevalence.
D. Incorrect. Colon cancer ranking 8 for both.

Question 183: A

The government has claimed a 20% reduction, so we are looking for an assessment criterion which has reduced 20% from 2013 to 2014. We can see that only "Number of people waiting for over 4 hours in A&E" has reduced by 20%, so this must be the criterion the government has used to describe "waiting times in A&E". Thus, the answer is A.

Question 184: B

Rovers must have played 10 games overall as they played each other's team twice. They lost 9 games scoring no points and so must have won 1 game, which scores 3 points.

Question 185: A

To have finished between City and United, Athletic must have got between 23 and 25 points. Hence they must have got 24 points because no team got the same number of points as another. Athletic won 7 games which is 21 points, so they must have also got 3 points from drawing 3 games. This accounts for all 10 games they played, so they did not lose any games.

Question 186: C

United won 8 games and drew 1, which is 25 points. Rangers drew 2 games and won none, which is 2 points. Therefore the difference in points is 23.

Question 187: C

Type 1 departments reached the new target of 95% at least three times since it was introduced. All the other statements are correct.

Question 188: C

Total attendances in Q1 08-9: 5.0 million

Total attendances in Q1 04-5: 4.5 million

The difference = 0.5 million

0.5/5 x 100 = 10% increase

Question 189: C

There are 16 quarters in total since the new target came into effect.

4/16 = 0.25, so the target has been hit 25% of the time i.e. missed 75% of the time.

Question 190: C

Ranjna must leave Singapore by 20:00 to get to Bali by 22:00. The latest flight she can therefore get is the 19:00. Thus, she must arrive in Singapore by 17:00 (accounting 2 hours for the stopover). The flight from Manchester to Singapore takes 14 hours. Manchester is 8 hours behind Singapore so she must leave Manchester 22 hours before 17:00 on Wednesday i.e. by 19:00 on Wednesday. Thus, the latest flight she can get is the 18:00 on Wednesday.

Question 191. D

The 08:00 flight will arrive at Singapore for 22:00 on Monday (GMT) or 06:00 Tuesday Singapore time. She then needs a 2 hour stopover, so earliest connecting flight she can get is 08:30 on Tuesday. The flight lands in Bali at 10:30. She then spends 1 hour and 45 minutes getting to her destination – arriving at 12:15 Tuesday.

Question 192: C

A. Incorrect. The graph is about level, and certainly not the steepest gradient post 2007.
B. Incorrect. Although there has been a general decline, there are some blips of increased smoking.
C. Correct.
D. Incorrect. The smoking rate in men decreased from 51% in 1974 to 21% in 2010. Thus, it decreased by more than a half.
E. Incorrect. The percentage difference between men and women smokers has been minimal in the 21st century.

Question 193: D

For this type of question you will have to use trial and error after you've analysed the data pattern to find the correct answer. The quickest way to do this is to examine outliers to try and match them to data in the table e.g. the left-most point is an outlier for the X-axis but average for the y-axis. Also look for any duplicated results in the table and if they are present on the graph, e.g. Hannah and Alice weigh 68 kg but this can't be found on the graph.

Question 194: C

This is pretty straightforward; the point is at approximately 172-174 cm in height and 164 -166 cm in arm span. Matthew is the only student who fits these dimensions.

Question 195: C

This is straightforward – just label the diagram using the information in the text and it becomes obvious that C is the correct answer.

Question 196: C
Since we do not know whether they went to university or not, we must add the number of women with children who work and those who went to university, 2, to the number of women with children who work but did not go to university, 1 (2 + 1 = 3).

Question 197: C
To work this out we must add up all the numbers within the rectangle, 4 + 6 + 1 + 2 + 11 + 12 + 7 + 15 = 58

Question 198: E
Calculate the number of men + women who have children and work i.e. 11 + 5 + 2 = 18

Question 199: C
To solve this we must work out the total number of people who had children i.e. 3 + 6 + 5 + 11 + 1 + 2 =28. Then we work out the total number of people who went to university, but that do not also have children so that these are not counted twice: 13 + 12 = 25. Then we add these two numbers together, 28 + 25 = 53 and subtract the number of people who fell into both categories i.e. 53 - (5 + 11 + 2) = 35

Question 200: C
To work this out we must add up all the numbers outside the rectangle that also fall within both the circle and the square, which is 5.

Question 201: D & E
This question asks for identification of the blank space, which is the space within the triangle, the rectangle and the square i.e. indicating working women who went to university but did not have children. This also reveals non-working men who did not have children and did not go to university.

Question 202: C
The normal price of these items would be £18.50 (£8 + £7 + £3.50). However, with the 50% discount on meat products, the price in the sale for these items will be £9.25. Thus, Alfred would receive £10.75 of change from a £20 note.

Question 203: C
The number of games played and points scored is a red herring in this question. The important data is 'Goals For' and 'Goals Against'. As this is a defined league and the teams have only played each other, the 'Goals For' column must equal the 'Goals Against' column.

Total Goals For = 16 + 11 + 8 + 7 + 8 + 4 = 54

Total Goals Against = 2 + Wilmslow + 7 + 9 + 12 + 14 = 44 + Wilmslow

For both columns to be equal, Wilmslow must have a total of 54 – 44 = 10 Goals Against.

Question 204: C
Working with the table it is possible to work out that the BMIs of Julie and Lydia must be 21 and 23, and hence their weights 100 and 115 lbs. Thus Emma's weight is 120 lbs, and her BMI must be 22, making her height equivalent to 160 cm.

Question 205: C
Working through the results, starting with the highest and lowest values, it is possible to plot all values and decipher which point is marked.

Question 206: D
This is a question of estimation. The average production across the year is at least 7 million barrels per day. Multiplying this by 365 gives around 2,550 million barrels per year. All other options require less than 7 million barrels daily production to be produced, and it is clear there is at least 7 million barrels per day. Therefore the answer is 2,700 million.

Alternatively we can estimate using 30 days per month, and multiplying the amount of barrels produced per day in each month by 30 (this is more accurate but more time consuming). 6+7+7+7.5+7.5+7+7.5+8+8.5+8.5+8+9 = 91.5, multiplying by 30 gives just over 2,700 million barrels.

Question 207: C

Use both graphs. For July, multiply the oil price by the amount sold in the month, and multiply by the number of days in the month. Thus, July = 7.5 million barrels x $75 per barrel x 31 days = $17,400 million = $17.4 billion

Question 208: B

Each three-block combination is mutually exclusive to any other combination, so the probabilities are added. Each block pick is independent of all other picks, so the probabilities can be multiplied. For this scenario there are three possible combinations:

P(2 red blocks and 1 yellow block) = P(red then red then yellow) + P(red then yellow then red) + P(yellow then red then red) =

$$(\frac{12}{20} \text{ x} \frac{11}{19} \text{ x} \frac{8}{18}) + (\frac{12}{20} \text{ x} \frac{8}{19} \text{ x} \frac{11}{18}) + (\frac{8}{20} \text{ x} \frac{12}{19} \text{ x} \frac{11}{18}) =$$

$$\frac{3 \text{ x } 12 \text{ x } 11 \text{ x } 8}{20 \text{ x } 19 \text{ x } 18} = \frac{44}{95}$$

Question 209: C

Multiply through by 15: $3(3x + 5) + 5(2x - 2) = 18 \text{ x } 15$

Thus: $9x + 15 + 10x - 10 = 270$

$9x + 10x = 270 - 15 + 10$

$19x = 265$

$x = 13.95$

Question 210: C

This is a rare case where you need to factorise a complex polynomial:

(3x)(x) = 0, possible pairs: 2 x 10, 10 x 2, 4 x 5, 5 x 4

$(3x - 4)(x + 5) = 0$

$3x - 4 = 0$, so $x = \frac{4}{3}$

$x + 5 = 0$, so $x = -5$

Question 211: C

$$\frac{5(x - 4)}{(x + 2)(x - 4)} + \frac{3\,(x + 2)}{(x + 2)(x - 4)}$$

$$= \frac{5x - 20 + 3x + 6}{(x + 2)(x - 4)}$$

$$= \frac{8x - 14}{(x + 2)(x - 4)}$$

Question 212: E

$p \propto \sqrt[3]{q}$, so $p = k\sqrt[3]{q}$

$p = 12$ when $q = 27$ gives $12 = k\sqrt[3]{27}$, so $12 = 3k$ and $k = 4$

so $p = 4\sqrt[3]{q}$

Now $p = 24$:

$24 = 4\sqrt[3]{q}$, so $6 = \sqrt[3]{q}$ and $q = 6^3 = 216$

Question 213: A

8 x 9 = 72

8 = (4 x 2) = 2 x 2 x 2

9 = 3 x 3

$(2 \times 2 \times 2 \times 3 \times 3)^2 = 2 \times 2 \times 2 \times 2 \times 2 \times 2 \times 3 \times 3 \times 3 \times 3 = 2^6 \times 3^4$

Question 214: C

Note that 1.151 x 2 = 2.302.

Thus: $\frac{2 \, x \, 10^5 + 2 \, x \, 10^2}{10^{10}} = 2 \, x \, 10^{-5} + 2 \, x \, 10^{-8}$

$= 0.00002 + 0.00000002 = 0.00002002$

Question 215: E

$y^2 + ay + b$

$= (y + 2)^2 - 5 = y^2 + 4y + 4 - 5$

$= y^2 + 4y + 4 - 5 = y^2 + 4y - 1$

So a = 4 and y = -1

Question 216: E

Take $5(m + 4n)$ as a common factor to give: $\frac{4(m + 4n)}{5(m + 4n)} + \frac{5(m - 2n)}{5(m + 4n)}$

Simplify to give: $\frac{4m + 16n + 5m - 10n}{5(m + 4n)} = \frac{9m + 6n}{5(m + 4n)} = \frac{3(3m + 2n)}{5(m + 4n)}$

Question 217: C

$A \, \alpha \frac{1}{\sqrt{B}}$. Thus, $= \frac{k}{\sqrt{B}}$.

Substitute the values in to give: $4 = \frac{k}{\sqrt{25}}$.

Thus, $k = 20$.

Therefore, $A = \frac{20}{\sqrt{B}}$.

When B = 16, $A = \frac{20}{\sqrt{16}} = \frac{20}{4} = 5$

Question 218: E

Angles SVU and STU are opposites and add up to 180°, so STU = 91°

The angle of the centre of a circle is twice the angle at the circumference so SOU = 2 x 91° = 182°

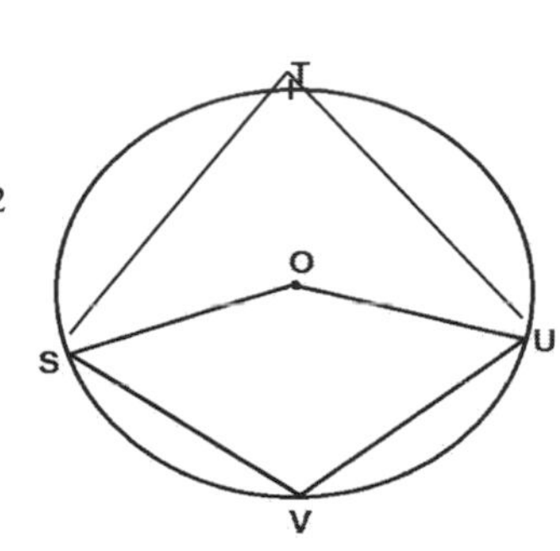

Question 219: E

The surface area of an open cylinder A = $2\pi rh$. Cylinder B is an enlargement of A, so the increases in radius (r) and height (h) will be proportional: $\frac{r_A}{r_B} = \frac{h_A}{h_B}$. Let us call the proportion coefficient n, where n = $\frac{r_A}{r_B} = \frac{h_A}{h_B}$.

So $\frac{Area\ A}{Area\ B} = \frac{2\pi r_A h_A}{2\pi r_B h_B} = n\ x\ n = n^2$. $\frac{Area\ A}{Area\ B} = \frac{32\pi}{8\pi} = 4$, so n = 2.

The proportion coefficient n = 2 also applies to their volumes, where the third dimension (also radius, i.e. the r^2 in $V = \pi r^2 h$) is equally subject to this constant of proportionality. The cylinder's volumes are related by $n^3 = 8$. If the smaller cylinder has volume 2π cm^3, then the larger will have volume $2\pi \times n^3 = 2\pi \times 8 = 16\pi$ cm^3.

Question 220: E

$$= \frac{8}{x(3-x)} - \frac{6(3-x)}{x(3-x)}$$

$$= \frac{8 - 18 + 6x}{x(3-x)}$$

$$= \frac{6x - 10}{x(3-x)}$$

Question 221: B

For the black ball to be drawn in the last round, white balls must be drawn every round. Thus the probability is given by $P = \frac{9}{10} \times \frac{8}{9} \times \frac{7}{8} \times \frac{6}{7} \times \frac{5}{6} \times \frac{4}{5} \times \frac{3}{4} \times \frac{2}{3} \times \frac{1}{2}$

$$= \frac{9 \times 8 \times 7 \times 6 \times 5 \times 4 \times 3 \times 2 \times 1}{10 \times 9 \times 8 \times 7 \times 6 \times 5 \times 4 \times 3 \times 2 \times 1} = \frac{1}{10}$$

Question 222: C

The probability of getting a king the first time is $\frac{4}{52} = \frac{1}{13}$, and the probability of getting a king the second time is $\frac{3}{51}$. These are independent events, thus, the probability of drawing two kings is $\frac{1}{13} x \frac{3}{51} = \frac{3}{663} = \frac{1}{221}$

Question 223: B

The probabilities of all outcomes must sum to one, so if the probability of rolling a 1 is x, then: $x + x + x + x + 2x = 1$. Therefore, $x = \frac{1}{7}$.

The probability of obtaining two sixes $P_{12} = \frac{2}{7} \times \frac{2}{7} = \frac{4}{49}$

Question 224: B

There are plenty of ways of counting, however the easiest is as follows: 0 is divisible by both 2 and 3. Half of the numbers from 1 to 36 are even (i.e. 18 of them). 3, 9, 15, 21, 27, 33 are the only numbers divisible by 3 that we've missed. There are 25 outcomes divisible by 2 or 3, out of 37.

Question 225: C

List the six ways of achieving this outcome: HHTT, HTHT, HTTH, TTHH, THTH and THHT. There are 2^4 possible outcomes for 4 consecutive coin flips, so the probability of two heads and two tails is: $6 x \frac{1}{2^4} = \frac{6}{16} = \frac{3}{8}$

	1	2	3	4	5	6
1	2	3	4	5	6	7
2	3	4	5	6	7	8
3	4	5	6	7	8	9
4	5	6	7	8	9	10
5	6	7	8	9	10	11
6	7	8	9	10	11	12

Question 226: D

Count the number of ways to get a 5, 6 or 7 (draw the square if helpful). The ways to get a 5 are: 1, 4; 2, 3; 3, 2; 4, 1. The ways to get a 6 are: 1, 5; 2, 4; 3, 3; 4, 2; 5, 1. The ways to get a 7 are: 1, 6; 2, 5; 3, 4; 4, 3; 5, 2; 6, 1. That is 15 out of 36 possible outcomes.

Question 227: C

There are x+y+z balls in the bag, and the probability of picking a red ball is $\frac{x}{(x+y+z)}$ and the probability of picking a green ball is $\frac{z}{(x+y+z)}$. These are independent events, so the probability of picking red then green is $\frac{xz}{(x+y+z)^2}$ and the probability of picking green then red is the same. These outcomes are mutually exclusive, so are added.

Question 228: B

There are two ways of doing it, pulling out a red ball then a blue ball, or pulling out a blue ball and then a red ball. Let us work out the probability of the first: $\frac{x}{(x+y+z)} \times \frac{y}{x+y+z-1}$, and the probability of the second option will be the same. These are mutually exclusive options, so the probabilities may be summed.

Question 229: A

[x: Player 1 wins point, y: Player 2 wins point]

Player 1 wins in five rounds if we get: yxxxx, xyxxx, xxyxx, xxxyx.

(Note the case of xxxxy would lead to player 1 winning in 4 rounds, which the question forbids.)

Each of these have a probability of $p^4(1-p)$. Thus, the solution is $4p^4(1-p)$.

Question 230: F

$4x + 7 + 18x + 20 = 14$

$22x + 27 = 14$

Thus, $22x = -13$

Giving $x = -\frac{13}{22}$

Question 231: D

$r^3 = \frac{3V}{4\pi}$

Thus, $r = \left(\frac{3V}{4\pi}\right)^{1/3}$

Therefore, $S = 4\pi\left[\left(\frac{3V}{4\pi}\right)^{\frac{1}{3}}\right]^2 = 4\pi\left(\frac{3V}{4\pi}\right)^{\frac{2}{3}}$

$= \frac{4\pi(3V)^{\frac{2}{3}}}{(4\pi)^{\frac{2}{3}}} = (3V)^{\frac{2}{3}} \text{ x } \frac{(4\pi)^1}{(4\pi)^{\frac{2}{3}}}$

$= (3V)^{\frac{2}{3}}(4\pi)^{1-\frac{2}{3}} = (4\pi)^{\frac{1}{3}}(3V)^{\frac{2}{3}}$

Question 232: A

Let each unit length be x.

Thus, $S = 6x^2$. Therefore, $x = \left(\frac{S}{6}\right)^{\frac{1}{2}}$

$V = x^3$. Thus, $V = [\left(\frac{S}{6}\right)^{\frac{1}{2}}]^3$ so $V = \left(\frac{S}{6}\right)^{\frac{3}{2}}$

Question 233: B

Multiplying the second equation by 2 we get $4x + 16y = 24$. Subtracting the first equation from this we get $13y = 17$, so $y = \frac{17}{13}$. Then solving for x we get $x = \frac{10}{13}$. You could also try substituting possible solutions one by one, although given that the equations are both linear and contain easy numbers, it is quicker to solve them algebraically.

Question 234: A

Multiply by the denominator to give: $(7x + 10) = (3y^2 + 2)(9x + 5)$

Partially expand brackets on right side: $(7x + 10) = 9x\left(3y^2 + 2\right) + 5\left(3y^2 + 2\right)$

Take x terms across to left side: $7x - 9x\left(3y^2 + 2\right) = 5\left(3y^2 + 2\right) - 10$

Take x outside the brackets: $x[7 - 9\left(3y^2 + 2\right)] = 5\left(3y^2 + 2\right) - 10$

Thus: $x = \frac{5\left(3y^2 + 2\right) - 10}{7 - 9\left(3y^2 + 2\right)}$

Simplify to give: $x = \frac{\left(15y^2\right)}{\left(7 - 9\left(3y^2 + 2\right)\right)}$

Question 235: F

$$3x\left(\frac{3x^7}{x^{\frac{1}{3}}}\right)^3 = 3x\left(\frac{3^3x^{21}}{x^{\frac{3}{3}}}\right)$$

$$= 3x\,\frac{27x^{21}}{x} = 81x^{21}$$

Question 236: D

$$2x[2^{\frac{7}{14}}\,x^{\frac{7}{14}}] = 2x[2^{\frac{1}{2}}\,x^{\frac{1}{2}}]$$

$$= 2x(\sqrt{2}\,\sqrt{x}) = 2\left[\sqrt{x}\sqrt{x}\right][\sqrt{2}\,\sqrt{x}]$$

$$= 2\sqrt{2x^3}$$

Question 237: A

$A = \pi r^2$, therefore $10\pi = \pi r^2$

Thus, $r = \sqrt{10}$

Therefore, the circumference is $2\pi\sqrt{10}$

Question 238: D

$3.4 = 12 + (3 + 4) = 19$

$19.5 = 95 + (19 + 5) = 119$

Question 239: D

$$2.3 = \frac{2^3}{2} = 4$$

$$4.2 = \frac{4^2}{4} = 4$$

Question 240: F

This is a tricky question that requires you to know how to 'complete the square':

$(x + 1.5)(x + 1.5) = x^2 + 3x + 2.25$

Thus, $(x + 1.5)^2 - 7.25 = x^2 + 3x - 5 = 0$

Therefore, $(x + 1.5)^2 = 7.25 = \frac{29}{4}$

Thus, $x + 1.5 = \sqrt{\frac{29}{4}}$

Thus $x = -\frac{3}{2} \pm \sqrt{\frac{29}{4}} = -\frac{3}{2} \pm \frac{\sqrt{29}}{2}$

Question 241: B

Whilst you definitely need to solve this graphically, it is necessary to complete the square for the first equation to allow you to draw it more easily:

$(x + 2)^2 = x^2 + 4x + 4$

Thus, $y = (x + 2)^2 + 10 = x^2 + 4x + 14$

This is now an easy curve to draw ($y = x^2$ that has moved 2 units left and 10 units up). The turning point of this quadratic is to the left and well above anything in x^3, so the only solution is the first intersection of the two curves in the upper right quadrant around (3.4, 39).

Question 242: C

The easiest way to solve this is to sketch them (don't waste time solving them algebraically). As soon as you've done this, it'll be very obvious that $y = 2$ and $y = 1-x^2$ don't intersect, since the latter has its turning point at (0, 1) and zero points at $x = -1$ and 1. $y = x$ and $y = x^2$ intersect at the origin and (1, 1), and $y = 2$ runs through both.

Question 243: B

Notice that you're not required to get the actual values – just the number's magnitude. Thus, 897653 can be approximated to 900,000 and 0.009764 to 0.01. Therefore, 900,000 x 0.01 = 9,000

Question 244: C

Multiply through by 70: $7(7x + 3) + 10(3x + 1) = 14 \times 70$

Simplify: $49x + 21 + 30x + 10 = 980$

$79x + 31 = 980$

$$x = \frac{949}{79}$$

Question 245: A

Split the equilateral triangle into 2 right-angled triangles and apply Pythagoras' theorem:

$x^2 = \left(\frac{x}{2}\right)^2 + h^2$. Thus $h^2 = \frac{3}{4}x^2$

$h = \sqrt{\frac{3x^2}{4}} = \frac{\sqrt{3x^2}}{2}$

The area of a triangle = ½ x base x height $= \frac{1}{2}x\frac{\sqrt{3x^2}}{2}$

Simplifying gives: $x\frac{\sqrt{3x^2}}{4} = x\frac{\sqrt{3}\sqrt{x^2}}{4} = \frac{x^2\sqrt{3}}{4}$

Question 246: A

This is a question testing your ability to spot 'the difference between two squares'.

Factorise to give: $3 - \frac{7x(5x - 1)(5x + 1)}{(7x)^2(5x + 1)}$

Cancel out: $3 - \frac{(5x - 1)}{7x}$

Question 247: C

The easiest way to do this is to 'complete the square':

$(x - 5)^2 = x^2 - 10x + 25$

Thus, $(x - 5)^2 - 125 = x^2 - 10x - 100 = 0$

Therefore, $(x - 5)^2 = 125$

$x - 5 = \pm\sqrt{125} = \pm\sqrt{25}\sqrt{5} = \pm 5\sqrt{5}$

$x = 5 \pm 5\sqrt{5}$

Question 248: B

Factorise by completing the square:

$x^2 - 4x + 7 = (x - 2)^2 + 3$

Simplify: $(x - 2)^2 = y^3 + 2 - 3$

$x - 2 = \pm\sqrt{y^3 - 1}$

$x = 2 \pm \sqrt{y^3 - 1}$

Question 249: D

Square both sides to give: $(3x + 2)^2 = 7x^2 + 2x + y$

Thus: $y = (3x + 2)^2 - 7x^2 - 2x = \left(9x^2 + 12x + 4\right) - 7x^2 - 2x$

$y = 2x^2 + 10x + 4$

Question 250: C

This is a fourth order polynomial, which you aren't expected to be able to factorise at GCSE. This is where looking at the options makes your life a lot easier. In all of them, opening the bracket on the right side involves making $(y \pm 1)^4$ on the left side, i.e. the answers are hinting that $(y \pm 1)^4$ is the solution to the fourth order polynomial.

Since there are negative terms in the equations (e.g. $-4y^3$), the solution has to be:

$(y-1)^4 = y^4 - 4y^3 + 6y^2 - 4y + 1$

Therefore, $(y-1)^4 + 1 = x^5 + 7$

Thus, $y - 1 = \left(x^5 + 6\right)^{\frac{1}{4}}$

$y = 1 + \left(x^5 + 6\right)^{1/4}$

Question 251: A

Let the width of the television be 4x and the height of the television be 3x.

Then by Pythagoras: $(4x)^2 + (3x)^2 = 50^2$

Simplify: $25x^2 = 2500$

Thus: $x = 10$. Therefore: the screen is 30 inches by 40 inches, i.e. the area is 1,200 inches^2.

Question 252: C

Square both sides to give: $1 + \frac{3}{x^2} = (y^5 + 1)^2$

Multiply out: $\frac{3}{x^2} = (y^{10} + 2y^5 + 1) - 1$

Thus: $x^2 = \frac{3}{y^{10} + 2y^5}$

Therefore: $x = \sqrt{\frac{3}{y^{10} + 2y^5}}$

Question 253: C

The easiest way is to double the first equation and triple the second to get:

$6x - 10y = 20 \; and \; 6x + 6y = 39$.

Subtract the first from the second to give: $16y = 19$,

Therefore, $y = \frac{19}{16}$.

Substitute back into the first equation to give $x = \frac{85}{16}$.

Question 254: C

This is fairly straightforward; the first inequality is the easier one to work with: B and D and E violate it, so we just need to check A and C in the second inequality.

C: $1^3 - 2^2 < 3$, but A: $2^3 - 1^2 > 3$

Question 255: B

Whilst this can be done graphically, it's quicker to do algebraically (because the second equation is not as easy to sketch). Intersections occur where the curves have the same coordinates.

Thus: $x + 4 = 4x^2 + 5x + 5$

Simplify: $4x^2 + 4x + 1 = 0$

Factorise: $(2x + 1)(2x + 1) = 0$

Thus, the two graphs only intersect once at $x = -\frac{1}{2}$

Question 256: D

It's better to do this algebraically as the equations are easy to work with and you would need to sketch very accurately to get the answer. Intersections occur where the curves have the same coordinates. Thus: $x^3 = x$
$x^3 - x = 0$

Thus: $x\left(x^2 - 1\right) = 0$

Spot the 'difference between two squares': $x(x + 1)(x - 1) = 0$

Thus there are 3 intersections: at $x = 0, 1\ and - 1$

Question 257: E

Note that the line is the hypotenuse of a right angled triangle with one side unit length and one side of length ½.

By Pythagoras, $\left(\frac{1}{2}\right)^2 + 1^2 = x^2$

Thus, $x^2 = \frac{1}{4} + 1 = \frac{5}{4}$

$$x = \sqrt{\frac{5}{4}} = \frac{\sqrt{5}}{\sqrt{4}} = \frac{\sqrt{5}}{2}$$

Question 258: D

We can eliminate z from equation (1) and (2) by multiplying equation (1) by 3 and adding it to equation (2):

$3x + 3y - 3z = -3$ Equation (1) multiplied by 3

$\underline{2x - 2y + 3z = 8}$ Equation (2) then add both equations

$5x + y = 5$ We label this as equation (4)

Now we must eliminate the same variable z from another pair of equations by using equation (1) and (3):

$2x + 2y - 2z = -2$ Equation (1) multiplied by 2

$\underline{2x - y + 2z = 9}$ Equation (3) then add both equations

$4x + y = 7$ We label this as equation (5)

We now use both equations (4) and (5) to obtain the value of x:

$5x + y = 5$ Equation (4)

$\underline{-4x - y = -7}$ Equation (5) multiplied by -1

$x = -2$

Substitute x back in to calculate y:

$4x + y = 7$

$4(-2) + y = 7$

$-8 + y = 7$

$y = 15$

Substitute x and y back in to calculate z:

$x + y - z = -1$

$-2 + 15 - z = -1$

$13 - z = -1$

$-z = -14$

$z = 14$

Thus: $x = -2$, $y = 15$, $z = 14$

Question 259: D

This is one of the easier maths questions. Take 3a as a factor to give:

$3a(a^2 - 10a + 25) = 3a(a - 5)(a - 5) = 3a(a - 5)^2$

Question 260: B

Note that 12 is the Lowest Common Multiple of 3 and 4. Thus:

-3 (4x + 3y) = -3 (48) Multiply each side by -3

4 (3x + 2y) = 4 (34) Multiply each side by 4

-12x – 9y = -144

12x + 8y = 136 Add together

-y = -8

y = 8

Substitute y back in:

4x + 3y = 48

4x + 3(8) = 48

4x + 24 = 48

4x = 24

x = 6

Question 261: E

Don't be fooled, this is an easy question, just obey BODMAS and don't skip steps.

$$\frac{-(25-28)^2}{-36+14} = \frac{-(-3)^2}{-22}$$

This gives: $\frac{-(9)}{-22} = \frac{9}{22}$

Question 262: E

Since there are 26 possible letters for each of the 3 letters in the license plate, and there are 10 possible numbers (0-9) for each of the 3 numbers in the same plate, then the number of license plates would be:

(26) x (26) x (26) x (10) x (10) x (10) $= 17{,}576{,}000$

Question 263: B

Expand the brackets to give: $4x^2 - 12x + 9 = 0$.

Factorise: $(2x - 3)(2x - 3) = 0$.

Thus, only one solution exists, x = 1.5.

Note that you could also use the fact that the discriminant, $b^2 - 4ac = 0$ to get the answer.

Question 264: C

$$= \left(x^{\frac{1}{2}}\right)^{\frac{1}{2}} \left(y^{-3}\right)^{\frac{1}{2}}$$

$$= x^{\frac{1}{4}} y^{-\frac{3}{2}} = \frac{x^{\frac{1}{4}}}{y^{\frac{3}{2}}}$$

Question 265: A

Let x, y, and z represent the rent for the 1-bedroom, 2-bedroom, and 3-bedroom flats, respectively. We can write 3 different equations: 1 for the rent, 1 for the repairs, and the last one for the statement that the 3-bedroom unit costs twice as much as the 1-bedroom unit.

(1) $x + y + z = 1240$

(2) $0.1x + 0.2y + 0.3z = 276$

(3) $z = 2x$

Substitute $z = 2x$ in both of the two other equations to eliminate z:

(4) $x + y + 2x = 3x + y = 1240$

(5) $0.1x + 0.2y + 0.3(2x) = 0.7x + 0.2y = 276$

$-2(3x + y) = -2(1240)$ Multiply each side of (4) by -2

$10(0.7x + 0.2y) = 10(276)$ Multiply each side of (5) by 10

(6) $-6x -2y = -2480$ Add these 2 equations

(7) $7x + 2y = 2760$

$x = 280$

$z = 2(280) = 560$ Because $z = 2x$

$280 + y + 560 = 1240$ Because $x + y + z = 1240$

$y = 400$

Thus the units rent for £ 280, £ 400, £ 560 per week respectively.

Question 266: C

Following BODMAS:

$$= 5\left[5\left(6^2 - 5 \times 3\right) + 400^{\frac{1}{2}}\right]^{1/3} + 7$$

$$= 5\,[5(36 - 15) + 20]^{\frac{1}{3}} + 7$$

$$= 5\,[5(21) + 20]^{\frac{1}{3}} + 7$$

$$= 5\,(105 + 20)^{\frac{1}{3}} + 7$$

$$= 5\,(125)^{\frac{1}{3}} + 7$$

$$= 5\,(5) + 7$$

$$= 25 + 7 = 32$$

Question 267: B

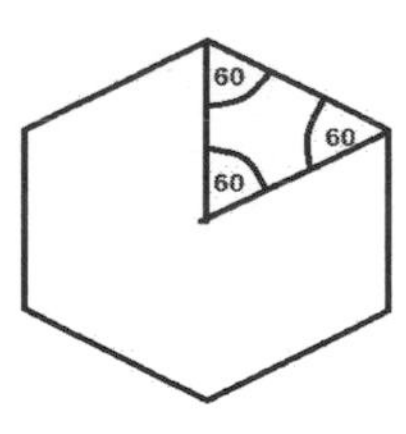

Consider a triangle formed by joining the centre to two adjacent vertices. Six similar triangles can be made around the centre – thus, the central angle is 60 degrees. Since the two lines forming the triangle are of equal length, we have 6 identical equilateral triangles in the hexagon.

Now split the triangle in half and apply Pythagoras' theorem:

$1^2 = 0.5^2 + h^2$

Thus, $h = \sqrt{\frac{3}{4}} = \frac{\sqrt{3}}{2}$

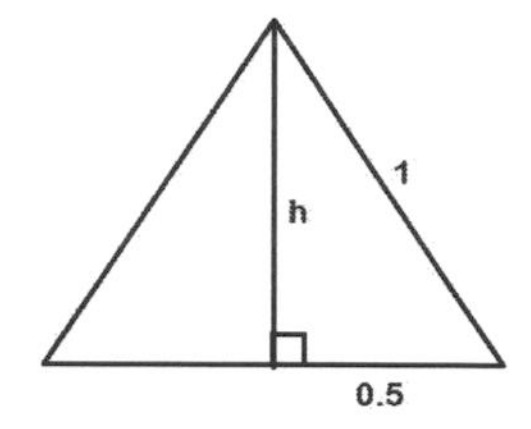

Thus, the area of the triangle is: $\frac{1}{2}bh = \frac{1}{2} \text{ x } 1 \text{ x} \frac{\sqrt{3}}{2} = \frac{\sqrt{3}}{4}$

Therefore, the area of the hexagon is: $\frac{\sqrt{3}}{4} \text{ x } 6 = \frac{3\sqrt{3}}{2}$

Question 268: B

Let x be the width and x+19 be the length.

Thus, the area of a rectangle is $x(x + 19) = 780$.

Therefore:

$x^2 + 19x – 780 = 0$

$(x - 20)(x + 39) = 0$

$x – 20 = 0$ or $x + 39 = 0$

$x = 20$ or $x = -39$

Since length can never be a negative number, we disregard $x = -39$ and use $x = 20$ instead.

Thus, the width is 20 metres and the length is 39 metres.

Question 269: B

The quickest way to solve is by trial and error, substituting the provided options. However, if you're keen to do this algebraically, you can do the following:

Start by setting up the equations: Perimeter = $2L + 2W = 34$

Thus: $L + W = 17$

Using Pythagoras: $L^2 + W^2 = 13^2$

Since $L + W = 17$, $W = 17 - L$

Therefore: $L^2 + (17 – L)^2 = 169$

$L^2 + 289 – 34L + L^2 = 169$

$2L^2 – 34L + 120 = 0$

$L^2 – 17L + 60 = 0$

$(L – 5) (L – 12) = 0$

Thus: $L = 5$ and $L = 12$

And: $W = 12$ and $W = 5$

Question 270: C

Multiply both sides by 8: $4(3x - 5) + 2(x + 5) = 8(x + 1)$
Remove brackets: $12x - 20 + 2x + 10 = 8x + 8$
Simplify: $14x - 10 = 8x + 8$
Add 10: $14x = 8x + 18$
Subtract 8x: $6x = 18$
Therefore: $x = 3$

Question 271: C

Recognise that 1.742 x 3 is 5.226. Now, the original equation simplifies to: $= \frac{3 \times 10^6 + 3 \times 10^5}{10^{10}}$

$= 3 \times 10^{-4} + 3 \times 10^{-5} = 3.3 \times 10^{-4}$

Question 272: A

$$Area = \frac{(2 + \sqrt{2})(4 - \sqrt{2})}{2}$$

$$= \frac{8 - 2\sqrt{2} + 4\sqrt{2} - 2}{2}$$

$$\frac{6 + 2\sqrt{2}}{2}$$

$$= 3 + \sqrt{2}$$

Question 273: C

Square both sides: $\frac{4}{x} + 9 = (y-2)^2$

$$\frac{4}{x} = (y-2)^2 - 9$$

Cross Multiply: $\frac{x}{4} = \frac{1}{(y-2)^2 - 9}$

$$x = \frac{4}{y^2 - 4y + 4 - 9}$$

Factorise: $x = \frac{4}{y^2 - 4y - 5}$

$$x = \frac{4}{(y+1)(y-5)}$$

Question 274: D

Set up the equation: $5x - 5 = 0.5(6x + 2)$

$10x - 10 = 6x + 2$

$4x = 12$

$x = 3$

Question 275: C

Round numbers appropriately: $\frac{55 + (\frac{9}{4})^2}{\sqrt{900}} = \frac{55 + \frac{81}{16}}{30}$

81 rounds to 80 to give: $\frac{55 + 5}{30} = \frac{60}{30} = 2$

Question 276: D

There are three outcomes from choosing the type of cheese in the crust. For each of the additional toppings to possibly add, there are 2 outcomes: 1 to include and another not to include a certain topping, for each of the 7 toppings

Thus, the number of different kinds of pizza is: 3 x 2 x 2 x 2 x 2 x 2 x 2 x 2 = 3 x 2^7

= 3 x 128 = 384

Question 277: A

Although it is possible to do this algebraically, by far the easiest way is via trial and error. The clue that you shouldn't attempt it algebraically is the fact that rearranging the first equation to make x or y the subject leaves you with a difficult equation to work with (e.g. $x = \sqrt{1 - y^2}$) when you try to substitute in the second.

An exceptionally good student might notice that the equations are symmetric in x and y, i.e. the solution is when x = y. Thus $2x^2 = 1$ and $2x = \sqrt{2}$ which gives $\frac{\sqrt{2}}{2}$ as the answer.

Question 278: C

If two shapes are congruent, then they are the same size and shape. Thus, congruent objects can be rotations and mirror images of each other. The two triangles in E are indeed congruent (SAS). Congruent objects must, by definition, have the same angles.

Question 279: B

Rearrange the equation: $x^2 + x - 6 \geq 0$

Factorise: $(x + 3)(x - 2) \geq 0$

Remember that this is a quadratic inequality so requires a quick sketch to ensure you don't make a silly mistake with which way the sign is.

Thus, $y = 0$ when $x = 2$ and $x = -3$. y > 0 when $x > 2$ or $x < -3$.

Thus, the solution is: $x \leq -3 \, and \, x \geq 2$.

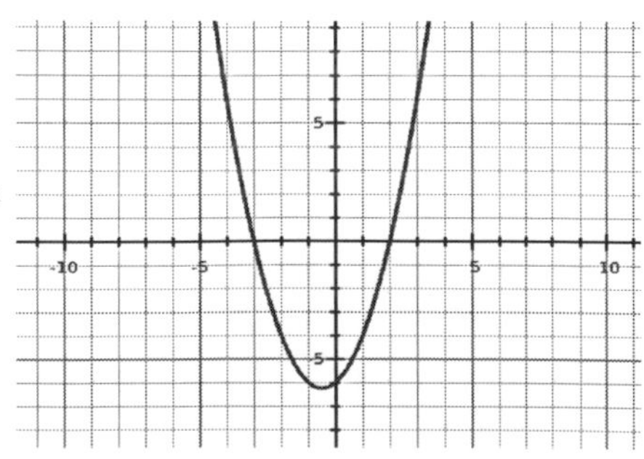

Question 280: B

Using Pythagoras: $a^2 + b^2 = x^2$

Since the triangle is equilateral: $a = b, \; so \; 2a^2 = x^2$

Area $= \frac{1}{2} base \, x \, height = \frac{1}{2} a^2$. From above, $a^2 = \frac{x^2}{2}$

Thus the area $= \frac{1}{2}x\frac{x^2}{2} = \frac{x^2}{4}$

Question 281: A

If X and Y are doubled, the value of Q increases by 4. Halving the value of A reduces this to 2. Finally, tripling the value of B reduces this to ⅔, i.e. the value decreases by ⅓.

Question 282: C

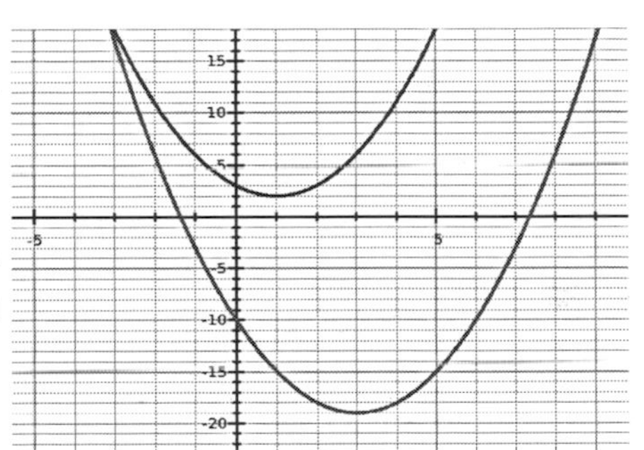

The quickest way to do this is to sketch the curves. This requires you to factorise both equations by completing the square:

$x^2 - 2x + 3 = (x-1)^2 + 2$

$x^2 - 6x - 10 = (x-3)^2 - 19$ Thus, the first equation has a turning point at (1, 2) and doesn't cross the x-axis. The second equation has a turning point at (3, -19) and crosses the x-axis twice.

Question 283: C

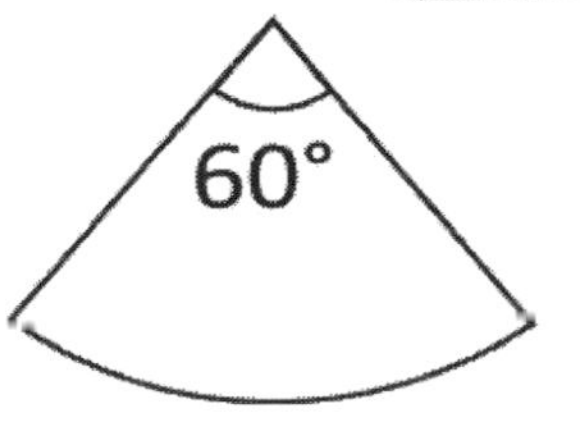

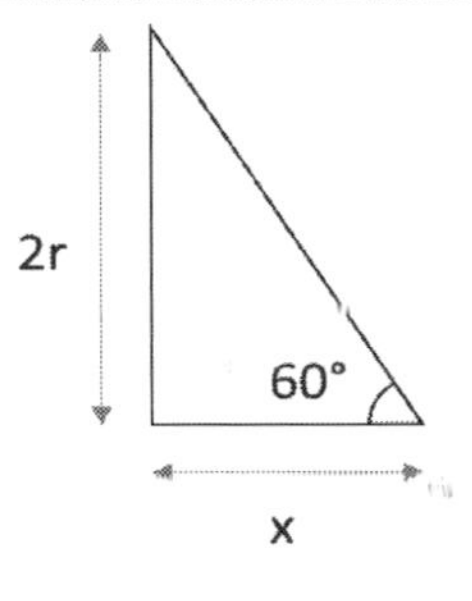

Segment area $= \frac{60}{360}\pi r^2 = \frac{1}{6}\pi r^2$

$\frac{x}{\sin 30°} = \frac{2r}{\sin 60°}$

$x = \frac{2r}{\sqrt{3}}$

Total triangle area $= 2 \times \frac{1}{2} \times \frac{2r}{\sqrt{3}} \times 2r = \frac{4r^2}{\sqrt{3}}$

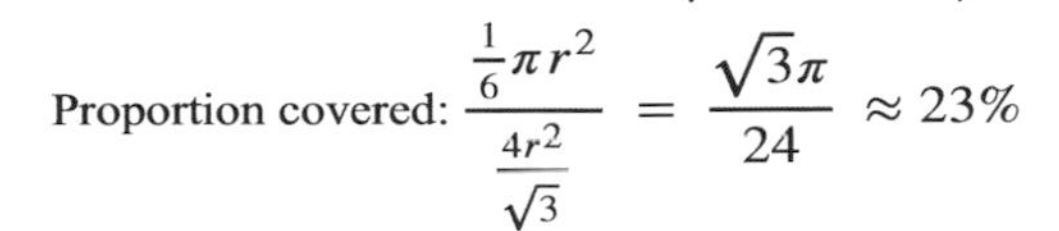

Proportion covered: $\frac{\frac{1}{6}\pi r^2}{\frac{4r^2}{\sqrt{3}}} = \frac{\sqrt{3}\pi}{24} \approx 23\%$

Question 284: B

$(2r)^2 = r^2 + x^2$

$3r^2 = x^2$

$x = \sqrt{3}r$

$Total\ height = 2r + x = (2+\sqrt{3})r$

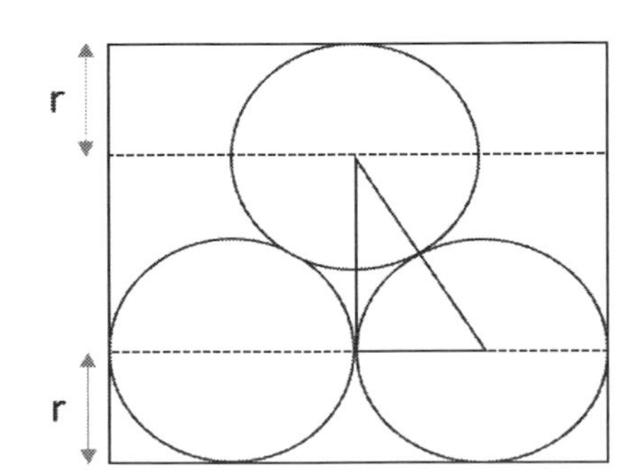

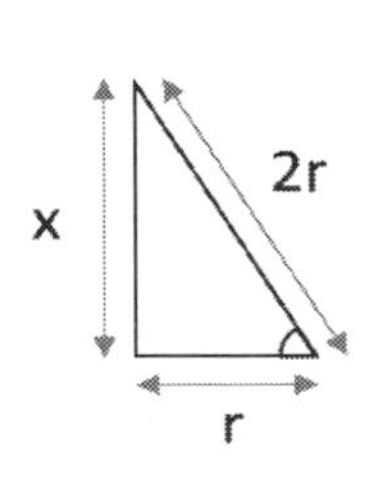

Question 285: A

$V = \frac{1}{3}h \times \text{base area}$

Therefore base area must be equal if h and V are the same

Internal angle = 180° – external ; external = 360°/6 = 60° giving internal angle 120°

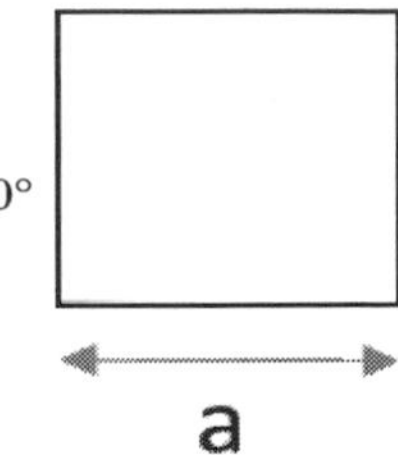

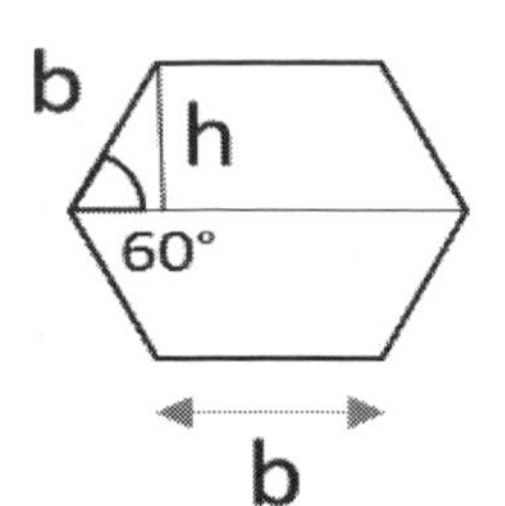

Hexagon is two trapezia of height h where: $\frac{b}{\sin 90°} = \frac{h}{\sin 60°}$

$$h = \frac{\sqrt{3}}{2}b$$

Trapezium area $= \frac{(2b + b)}{2}\frac{\sqrt{3}}{2}b = \frac{3\sqrt{3}}{4}b^2$

Total hexagon area $= \frac{3\sqrt{3}}{2}b^2$

So from equal volumes: $a^2 = \frac{3\sqrt{3}}{2}b^2$

Ratio: $\sqrt{\frac{3\sqrt{3}}{2}}$

Question 286: C

A cube has 6 sides so the area of 9 cm cube = 6 x 9^2
9 cm cube splits into 3 cm cubes.
Area of 3 cm cubes = 3^3 x 6 x 3^2

$$\frac{6 \times 3^2 \times 3^3}{6 \times 3^2 \times 3^2} = 3$$

Question 287: E

$$x^2 = (4r)^2 + r^2$$
$$x = \sqrt{17}r$$
$$\frac{\sqrt{17}r}{\sin 90°} = \frac{r}{\sin\theta}$$
$$\theta = \sin^{-1}\left(\frac{1}{\sqrt{17}}\right)$$

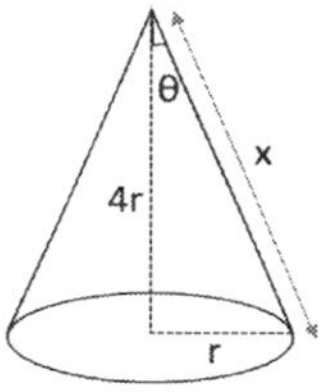

Question 288: C

0 to 200 is 180 degrees so: $\frac{\theta}{180} = \frac{70}{200}$

$$\theta = \frac{7 \times 180}{20} = 63°$$

Question 289: C

Since the rhombi are similar, the ratio of angles = 1
Length scales with square root of area so length B = $\sqrt{10}$ length A

$$\frac{\frac{angle\ A}{angle\ B}}{\frac{length\ A}{length\ B}} = \frac{1}{\frac{\sqrt{10}}{1}} = \frac{1}{\sqrt{10}}$$

Question 290: E

$$y = \ln(2x^2)$$
$$e^y = 2x^2$$

$x = \sqrt{\frac{e^y}{2}}$

As the input is -x, the inverse function must be $f(x) = -\sqrt{\frac{e^y}{2}}$

Question 291: C

$log_8(x)$ and $log_{10}(x) < 0$; $x^2 < 1$; $\sin(x) \leq 1$ and $1 < e^x < 2.72$

So e^x is largest over this range

Question 292: C

$x \propto \sqrt{z}^3$

$\sqrt{2}^3 = 2\sqrt{2}$

Question 293: A

The area of the shaded part, that is the difference between the area of the larger and smaller circles, is three times the area of the smaller so: $\pi r^2 - \pi x^2 = 3\pi x^2$. From this, we can see that the area of the larger circle, radius x, must be 4x the smaller one so: $4\pi r^2 = \pi x^2$

$4r^2 = x^2$

$x = 2r$

The gap is $x - r = 2r - r = r$

Question 294: D

$x^2 + 3x - 4 \geq 0$

$(x - 1)(x + 4) \geq 0$

Hence, $x - 1 \geq 0$ or $x + 4 \geq 0$

So $x \geq 1$ or $x \geq -4$

Question 295: C

$\frac{4}{3}\pi r^3 = \pi r^2$

$\frac{4}{3}r = 1$

$r = \frac{3}{4}$

Question 296: B

When $x^2 = \frac{1}{x}$; $x = 1$

When $x > 1, x^2 > 1, \frac{1}{x} < 1$

When $x < 1, x^2 < 1, \frac{1}{x} > 1$

Range for $\frac{1}{x}$ is $x > 0$

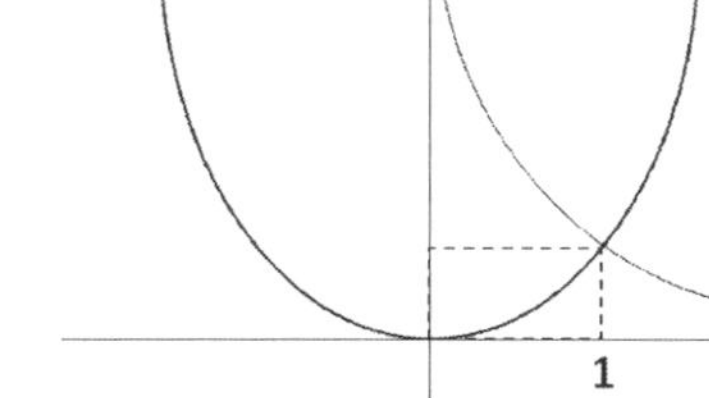

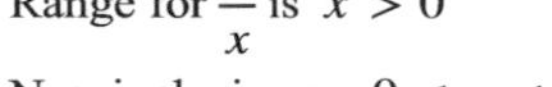

Non-inclusive so: $0 < x < 1$

Question 297: A

Don't be afraid of how difficult this initially looks. If you follow the pattern, you get (e-e) which = 0. Anything multiplied by 0 gives zero.

Question 298: C

For two vectors to be perpendicular their scalar product must be equal to 0.

Hence, $\begin{pmatrix} -1 \\ 6 \end{pmatrix} \cdot \begin{pmatrix} 2 \\ k \end{pmatrix} = 0$

$\therefore \; -2 + 6k = 0$

$k = \frac{1}{3}$

Question 299: C

The point, q, in the plane meets the perpendicular line from the plane to the point p.

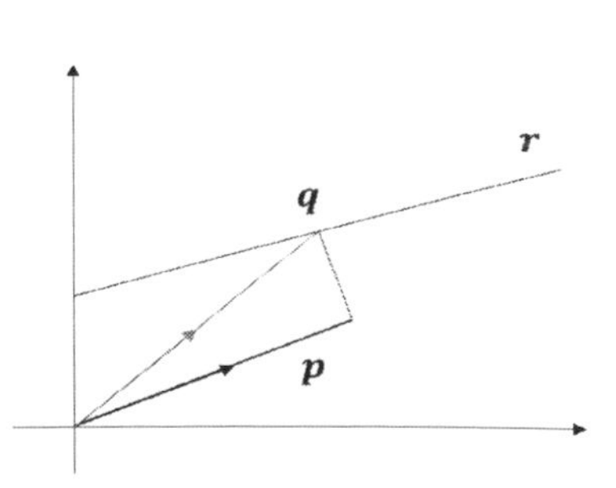

$\boldsymbol{q} = -3\boldsymbol{i} + \boldsymbol{j} + \lambda_1(\boldsymbol{i} + 2\boldsymbol{j})$

$\overrightarrow{PQ} = -3\boldsymbol{i} + \boldsymbol{j} + \lambda_1(\boldsymbol{i} + 2\boldsymbol{j}) + 4\boldsymbol{i} + 5\boldsymbol{j}$

$= \begin{pmatrix} -7 + \lambda_1 \\ -4 + 2\lambda_1 \end{pmatrix}$

PQ is perpendicular to the plane $\boldsymbol{r}$ therefore the dot product of $\overrightarrow{PQ}$ and a vector within the plane must be 0.

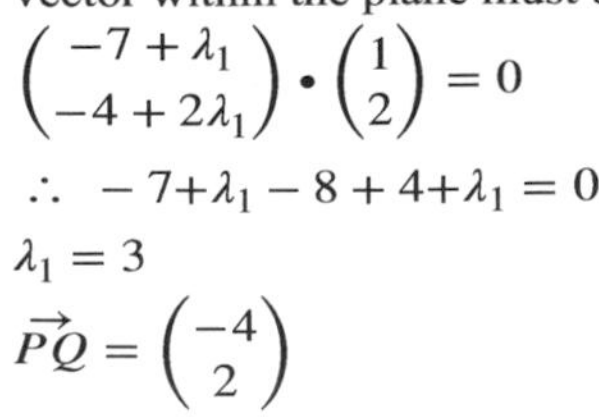

$\begin{pmatrix} -7 + \lambda_1 \\ -4 + 2\lambda_1 \end{pmatrix} \cdot \begin{pmatrix} 1 \\ 2 \end{pmatrix} = 0$

$\therefore \; -7 + \lambda_1 - 8 + 4 + \lambda_1 = 0$

$\lambda_1 = 3$

$\overrightarrow{PQ} = \begin{pmatrix} -4 \\ 2 \end{pmatrix}$

The perpendicular distance from the plane to point p is therefore the modulus of the vector joining the two $\overrightarrow{PQ}$:

$$\left|\overrightarrow{PQ}\right| = \sqrt{(-4)^2 + 2^2} = \sqrt{20} = 2\sqrt{5}$$

Question 300: E

$-1 + 3\mu = -7\ ;\ \mu = -2$
$2 + 4\lambda + 2\mu = 2\ \therefore\ \lambda = 1$
$3 + \lambda + \mu = k\ \therefore\ k = 2$

Question 301: E

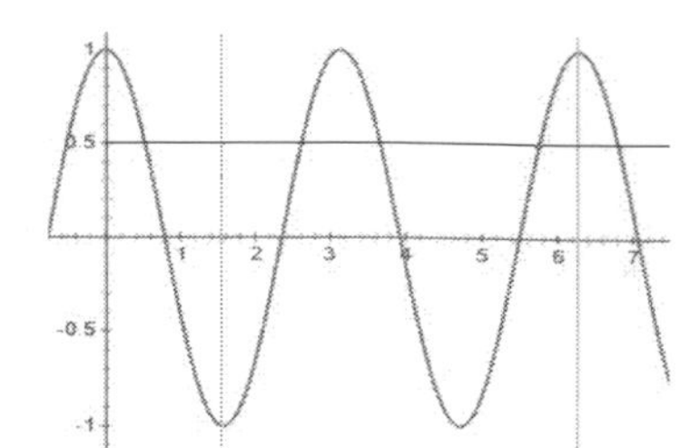

$\sin\left(\frac{\pi}{2} - 2\theta\right) = \cos(2\theta)$

Root solution to $\cos(\theta) = 0.5$
$\theta = \frac{\pi}{3}$

Solution to $\cos(2\theta) = 0.5$
$\theta = \frac{\pi}{6}$

Largest solution within range is: $2\pi - \frac{\pi}{6} = \frac{(12-1)\pi}{6} = \frac{11\pi}{6}$

Question 302: A

$\cos^4(x) - \sin^4(x) \equiv \left\{\cos^2(x) - \sin^2(x)\right\}\left\{\cos^2(x) + \sin^2(x)\right\}$
From difference of two squares, then using Pythagorean identity $\cos^2(x) + \sin^2(x) = 1$
$\cos^4(x) - \sin^4(x) \equiv \cos^2(x) - \sin^2(x)$
But double angle formula says: $\cos(A+B) = \cos(A)\cos(B) - \sin(A)\sin(B)$
$\therefore if A = B,\ \cos(2A) = \cos(A)\cos(A) - \sin(A)\sin(A)$
$= \cos^2(A) - \sin^2(A)$
So, $\cos^4(x) - \sin^4(x) \equiv \cos(2x)$

Question 303: C

Factorise: $(x+1)(x+2)(2x-1)\left(x^2+2\right) = 0$

Three real roots at $x = -1, x = -2, x = 0.5$ and two imaginary roots at 2i and -2i

Question 304: C

An arithmetic sequence has constant difference d so the sum increases by d more each time:
$u_n = u_1 + (n-1)d$
$\sum_1^n u_n = \frac{n}{2}\{2u_1 + (n-1)d\}$
$\sum_1^8 u_n = \frac{8}{2}\{4 + (8-1)3\} = 100$

Question 305: E

$\binom{n}{k}2^{n-k}(-x)^k = \binom{5}{2}2^{5-2}(-x)^2$
$= 10 \times 2^3x^2 = 80x^2$

Question 306: A

Having already thrown a 6 is irrelevant. A fair die has equal probability $P = \frac{1}{6}$ for every throw.

For three throws: $P(6 \cap 6 \cap 6) = \left(\frac{1}{6}\right)^3 = \frac{1}{216}$

Question 307: D

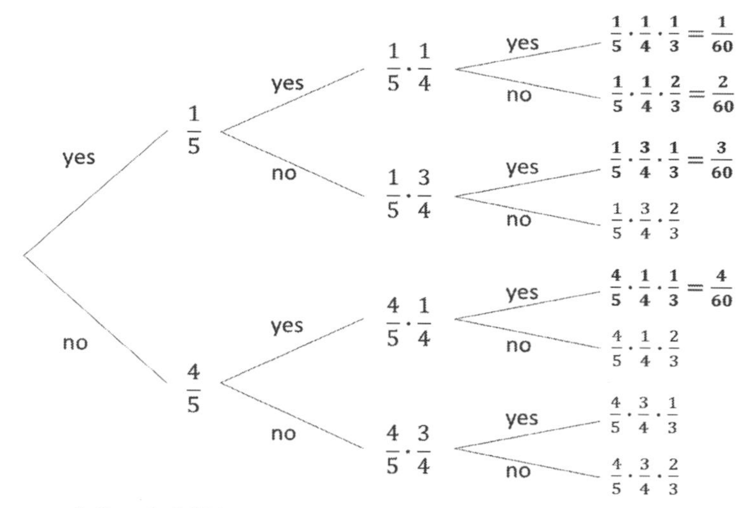

Total probability is sum of all probabilities:

$= P(Y \cap Y \cap Y) + P(Y \cap Y \cap N) + P(Y \cap N \cap Y) + P(N \cap Y \cap Y)$

$= \frac{1}{60} + \frac{2}{60} + \frac{3}{60} + \frac{4}{60} = \frac{10}{60} = \frac{1}{6}$

Question 308: C

$P\left[(A \cup B)'\right] = 1 - P\left[(A \cup B)\right]$

$= 1 - \{P(A) + P(B) - P(A \cap B)\}$

$= 1 - \frac{2+6-1}{8} = \frac{3}{8}$

Question 309: D

Using the product rule: $\frac{dy}{dx} = x \bullet 4(x+3)^3 + 1 \bullet (x+3)^4$

$= 4x(x+3)^3 + (x+3)(x+3)^3$

$= (5x+3)(x+3)^3$

Question 310: A

$\int_1^2 \frac{2}{x^2} dx = \int_1^2 2x^{-2} dx =$

$\left[\frac{2x^{-1}}{-1}\right]_1^2 = \left[\frac{-2}{x}\right]_1^2$

$= \frac{-2}{2} - \frac{-2}{1} = -1$

Question 311: D

Express $\frac{5i}{1+2i}$ in the form $a + bi$

$\frac{5i}{1+2i} \bullet \frac{1-2i}{1-2i}$

$= \frac{5i+10}{1+4} - \frac{5i+10}{5}$

$= i + 2$

Question 312: B

$7\log_a(2) - 3\log_a(12) + 5\log_a(3)$

$7\log_a(2) = \log_a\left(2^7\right) = \log_a(128)$

$3\log_a(12) = \log_a(1728)$

$5\log_a(3) = \log_a(243)$

This gives: $\log_a(128) - \log_a(1728) + \log_a(243)$

$$= \log_a\left(\frac{128 \times 243}{1728}\right) = \log_a(18)$$

Question 313: E

Functions of the form quadratic over quadratic have a horizontal asymptote.
Divide each term by the highest order in the polynomial i.e. x^2:

$$\frac{2x^2 - x + 3}{x^2 + x - 2} = \frac{2 - \frac{1}{x} + \frac{3}{x^2}}{1 + \frac{1}{x} - \frac{2}{x^2}}$$

$$\lim_{x \to \infty}\left(\frac{2 - \frac{1}{x} + \frac{3}{x^2}}{1 + \frac{1}{x} - \frac{2}{x^2}}\right) = \frac{2}{1} \quad i.e.\ y \to 2$$

So, the asymptote is $y = 2$

Question 314: A

$1 - 3e^{-x} = e^x - 3$

$$4 = e^x + 3e^{-x} = \frac{(e^x)^2}{e^x} + \frac{3}{e^x} = \frac{(e^x)^2 + 3}{e^x}$$

This is a quadratic equation in (e^x): $(e^x)^2 - 4(e^x) + 3 = 0$

$(e^x - 3)(e^x - 1) = 0$

So $e^x = 3, x = \ln(3)$ or $e^x = 1, x = 0$

Question 315: D

Rearrange into the format: $(x + a)^2 + (y + b)^2 = r^2$

$(x - 3)^2 + (y + 4)^2 - 25 = 12$

$(x - 3)^2 + (y + 4)^2 = 47$

$\therefore r = \sqrt{47}$

Question 316: C

$\sin(-x) = -\sin(x)$

$$\int_0^a 2\sin(-x)dx = -2\int_0^a \sin(x)dx = -2\left[\cos(x)\right]_0^a = \cos(a) - 1$$

Solve $\cos(a) - 1 = 0 \therefore a = 2k\pi$

Or simply the integral of any whole period of sin(x) = 0 i.e. $a = 2k\pi$

Question 317: E

$$\frac{2x + 3}{(x - 2)(x - 3)^2} = \frac{A}{(x - 2)} + \frac{B}{(x - 3)} + \frac{C}{(x - 3)^2}$$

$2x + 3 = A(x - 3)^2 + B(x - 2)(x - 3) + C(x - 2)$

When $x = 3$, $(x - 3) = 0$, $C = 9$

When $x = 2$, $(x - 2) = 0$, $A = 7$

$2x + 3 = 7(x - 3)^2 + B(x - 2)(x - 3) + 9(x - 2)$

For completeness: Equating coefficients of x^2on either side: $0 = 7 + B$ which gives: $B = -7$

END OF SECTION

Past Paper Worked Solutions

Hundreds of students take the ECAA exam each year. These exam papers are then released online to help future students prepare for the exam. Since the ECAA is such a new exam, past papers have become an invaluable resource in any student's preparation.

Where can I get ECAA Past Papers?

This book does not include ECAA past paper questions because it would be over 500 pages long if it did! However, all ECAA past papers since 2016 (including the specimen paper) are available for free from the official ECAA website. To save you the hassle of downloading lots of files, we've put them all into one easy-to-access folder for you at **www.uniadmissions.co.uk/ECAA-past-papers**.

How should I use ECAA Past Papers?

ECAA Past papers are one the best ways to prepare for the ECAA. Careful use of them can dramatically boost your scores in a short period of time. The way you use them will depend on your learning style and how much time you have until the exam date but here are some general pointers:

- Four to eight weeks of preparation is usually sufficient for most students.
- Make sure you are completely comfortable with the ECAA syllabus before attempting past papers – they are a scare resource and you shouldn't 'waste them' if you're not fully prepared to take them.
- Its best to start working through practice questions before tackling full papers under time conditions.
- You can find two additional mock papers in the *ECAA Practice Papers* Book (flick to the back to get a free copy).

How should I use past papers?

This book is designed to accelerate your learning from ECAA past papers. Avoid the urge to have this book open alongside a past paper you're seeing for the first time. The ECAA is difficult because of the intense time pressure it puts you under – the best way of replicating this is by doing past papers under strict exam conditions (no half measures!). Don't start out by doing past papers (see previous page) as this 'wastes' papers.

Once you've finished, take a break and then mark your answers. Then, review the questions that you got wrong followed by ones which you found tough/spent too much time on. This is the best way to learn and with practice, you should find yourself steadily improving. You should keep a track of your scores on the next page so you can track your progress.

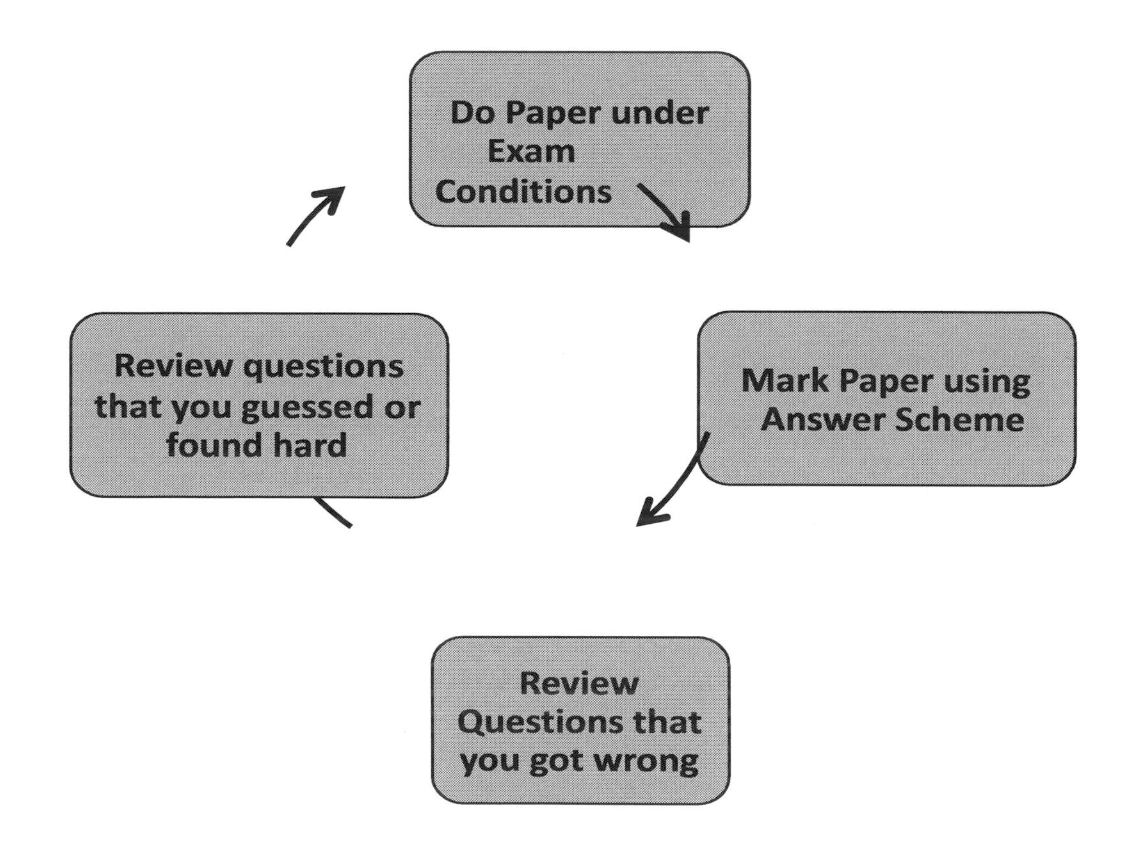

Scoring Tables

Use these to keep a record of your scores – you can then easily see which paper you should attempt next (always the one with the lowest score).

SECTION 1A	1st Attempt	2nd Attempt	3rd Attempt
Specimen	9/20		
2016	11/22	19/22	
2017	6/20	18/20	

SECTION 1B	1st Attempt	2nd Attempt	3rd Attempt
Specimen	5/16		
2016	9/15		
2017	5/16	7/16	

SECTION 2	1st Attempt	2nd Attempt	3rd Attempt
Specimen			
2016			
2017			

Specimen

Section 1A

Question 1: C

The answer is option C. The maximum number of days in a calendar month is 31. Each day of the week occurs four times during the first 28 days of every month. Two of these days will be working Saturdays. If the 29th, 30th and 31st of a 31-day month are all weekdays, or if the 31st is a working Saturday, the number of working days during the month will be $(4 \times 5) + 2 + 3 = 25$. In a month when the neighbours work the maximum 25 days, one of them will drive 12 times and the other one will drive 13 times.

Question 2: D

The journey to work is 1800m, which would take $1800 \div 5 = 360$ seconds = 6 minutes if there was no delay at either of the sets of lights. It takes $900 \div 5 = 180$ seconds = 3 minutes to cycle from the first set of lights to the second. Because both sets change simultaneously, the maximum wait of 2 minutes at one of them will mean a wait of 1 minute at the other. The longest journey time is therefore $6 + 2 + 1 = 9$ minutes.

Question 3: B

The answer is option B. There are eight possible side views of this paperweight: four with TLF uppermost (as shown in the question) and four with the reflection of TLF uppermost. You should visualise these views and eliminate the four options that are side views. From the side, a vertical line will be seen at each boundary between a projection and a recess. A and E are both views of the bases of the letters T, L and F. A is the view when TLF is uppermost and E is the view when the reflection of TLF is uppermost. C is the view of the tops of the letters T, L and F, with TLF uppermost. D is the view of the edge of the letter F, with the reflection of TLF uppermost. (Note that a view of the edge of the letter T would have a single vertical line one fifth of the way from one end.)

Question 4: A

The answer is option A. The energy values in the table are for 100g of each food. The energy provided by each of the first four ingredients is, therefore: mealworms – $1.5 \times 150 = 225$ calories apples – $1.5 \times 350 = 525$ calories raisins – $2.5 \times 300 = 750$ calories suet – $1.25 \times 800 = 1000$ calories The total energy provided by these ingredients is $225 + 525 + 750 + 1000 = 2500$ calories. This means that the sunflower seeds must provide 2500 calories in order to make up the required total of 5000 calories. Sunflower seeds provide 500 calories per 100g, so $(2500 \div 500) \times 100 = 500$ g of sunflower seeds must be used.

Question 5: B

The distance between Sue and Freya increases while Freya runs to collect the stick and decreases as she brings it back to Sue. Only options B and D show this situation. If Sue were to stand still, it would take the same amount of time for Freya to run 20m in both directions (assuming the same average speed both ways), but because Sue is walking towards her, the distance between them decreases from 20m to 0 in a shorter time than it increases from 0 to 20 m.

Question 6: E

Because Alistair changes the code each time, you must not allow yourself to be distracted by the symbols making up his name. You simply need to identify which one of the options has repeated letters in the positions of the repeated symbols in the reply. Only SOMETIME has the 7th letter the same as the 3rd letter and the 8th letter the same as the 4th letter.

Question 7: D

The two journeys took a total of 24/ 32 hours, which is 45 minutes. This means that on his second journey he travelled 24 – 15 = 9km in 45 – 30 = 15 minutes (¼ hour). Speed = distance ÷ time, so his average speed on the second journey was 9km ÷ ¼ hour = 36km/h.

Question 8: C

This question requires you to extract and process the relevant data from both the table and the narrative. You will need to compare the cost of silver membership for one year, which includes a free locker, with the cost of bronze membership for one year plus locker hire. The cost of silver membership for one year is the renewal price of £28. The cost of bronze membership for one year plus locker hire for 12 visits (one per month) is the renewal price of £8 + (12 × £2) = £32. The cheaper option is silver membership for £28, which is £12 less than the £40 paid for gold membership 6 months ago as a new member.

Question 9: A

To answer this question you first need to establish that the number of 11-year olds who go swimming (104) is slightly greater than 2 /3 of the number of 16-year olds who go swimming. It is not necessary to carry out exact calculations for any of the sports; you only need to observe the following: Football – 120 / 181 is very close to 2/3. Cricket – 120 / 133 is considerably greater than 2/3. Hockey – 55 / 66 is considerably greater than 2/3. Tennis – 123 / 149 is considerably greater than 2/3. Squash – 51 / 97 is slightly greater than 1/2.

Question 10: B

The answer is option B. The moving average of 13.82 beside 1913 means that x + 13.3 + 14.9 + 13.7 + 14.2 = 13.82 × 5, so x + 56.1 = 69.1 and x = 13.0. y is therefore (13.4 + 14.4 + 13.4 + 13.7 + 13.0) ÷ 5 = 67.9 ÷ 5 = 13.58

Question 11: B

A group of 10 students would only require 2 instructors, who would be paid £12 each for 2 hours, but the extra £12 for a third instructor is much less than the extra income from a group of 18 students. A group of 18 students would pay a total of 18 × (£7 + £5) = £216 for 2 hours, so the maximum profit the centre can make from a group after paying the instructors for a two-hour session is £216 – £36 = £180.

Question 12: E

Grace has been late to lessons more than twice, so she can be eliminated. Andrew and Edward have both failed to complete more than two pieces of homework by the deadline set, so they can also be eliminated. Carole has 3 non-A-grade pieces of work, but Ian has only 2 pieces of non-A-grade work, so Ian will be awarded the prize.

Question 13: B

Because each group involves the same number of people, this question can be approached by assuming that 400 people were surveyed altogether (100 in each age group) and then analysing the five statements as follows: In the age group 25 – 35, 32 people preferred Drink A and 33 people preferred Drink B, so conclusion A cannot be drawn. A total of 103 people preferred Drink B and 101 people had no preference, so conclusion B can be drawn. A total of 48 people in the age group 5 – 15 expressed a preference. This is less than half of the people in the group, so conclusion C cannot be drawn. A total of 225 people had a preference. This is more than half of the total number of people surveyed, so conclusion D cannot be drawn. A total of 74 people didn't know which they preferred. 20% of 400 = 80, so conclusion E cannot be drawn.

Question 14: D

The fact that the bills for April and November are different amounts for the same number of units used means that the original tariff applies to at least the first four months of the year. From January to February (for instance) an increase of 25 units used increased the monthly charge by £2.50, so the original charge per unit was 10p per unit. From January to August the formula: £30 + (units used) × 10p consistently gives the monthly charge, but in September it doesn't. (The tariff changed to £20 standing charge per month plus 20p per unit used in September, but it is not necessary to calculate this in order to answer the question.)

Question 15: C

The most efficient way of approaching this question is to solve the problem numerically. However, because it is the method of calculation that is important here and not the actual figures, time can be saved by using any number greater than 420 for the total number of desks available. The total number of desks available in the 12 rooms (including the sports hall, though it doesn't make any difference to the method if you don't include it) is 456, so there are 36 spare desks (456 – 420). The exams officer wants to leave the same number of empty desks in each room, so he should divide 36 by 12 and subtract the answer (3) from the number of desks in each room. This is the method described in option C.

Question 16: D

The most efficient way of approaching this question is to work out your own answer from the graph and then select the option that is the closest. A reasonable estimate for the cumulative rainfall at the beginning of June is 175 mm (three quarters of the way from 100mm to 200 mm) and it is clearly 400 mm at the end of September. This gives an average of 225 ÷ 4 = 56.25 mm /month for the 4-month period. This is very close to 57mm/month, which is option D.

Question 17: B

The hot tap supplies 24 litres of water per minute (360 litres ÷ 15 minutes) and the cold tap supplies 36 litres of water per minute (360 litres ÷ 10 minutes). Together they supply a total of 90 litres in the 1½ minutes before the cold tap is turned off, which makes the bath ¼ full. The hot tap then supplies another 180 litres to make the bath ¾ full, taking a further 180 ÷ 24 = 7½ minutes. In all five options the graph rises from (0, 0) to (1, 1) as the two taps together make the bath ¼ full in the first 1½ minutes. The correct graph then has to rise 2 units up the y-axis (from ¼ to ¾) as it advances 5 units along the x-axis (7½ minutes).

Question 18: E

The reflection of 22:05, and therefore the correct time when the watch was reset to 22:05, is 20:55. This means that the watch is 1 hour 10 minutes ahead of the correct time. As a result, when the watch displayed the time as 10:09 this morning, the view of the digital clock that could be seen was the reflection of 08:59.

Question 19: A

The bar chart reveals that the usage this year will be as follows (assuming it to be the same as last year, as instructed); January – March: 6,000; April – June: 7,000; July – September: 5,000; October – December: 6,000. There were 2,000 containers left at the end of last year, so it would appear that only 4,000 need to be delivered in January. However, this would mean that there would be none left at the end of March, and, because the maximum number that can be delivered at one time is 6,000, they would run out before the end of June. Ordering 5,000 in January would raise the total in stock to 7,000, with 1,000 left at the end of March. If 6,000 were then to be ordered in April there would be just enough to cover the usage of 7,000 from April to June. Following this, 5,000 could be ordered in July and 6,000 in October. This would result in there being none leftover at the end of the year.

Question 20: E

To answer this question you need to visualize the appearances of the trees from different parts of field Y. The most efficient approach is to imagine walking from one side of the field to the other. Starting from the extreme top left (for example), the order of the trees seen changes as follows: 4, 3, 2, 1 4, 1 (with 3 behind), 2 4, 1, 3, 2 1 (with 4 behind), 3, 2 1, 4, 3, 2 1, 4, 2 (with 3 behind) 1, 4, 2, 3 1, 2 (with 4 behind), 3 1, 2, 4, 3 Photographs were only taken when all four trees were visible, so there can be five different orders in the photographs.

Section 1B

Question 21: B
The answer is option B.

Use the product rule, such that:

$$u = (3x-2)^2,\ u' = 2(3x-2)(2)v = x^{-\frac{3}{2}},\ v' = \left(-\frac{3}{2}\right)x^{-\frac{5}{2}}$$

$$\frac{dy}{dx} = (3x-2)^2\left(-\frac{3}{2}x^{-\frac{5}{2}}\right) + \left(x^{-\frac{3}{2}}\right)(6)(3x-2)$$

Substituting x = 2 into this equation, we eventually find that

$$\frac{dy}{dx} = (4^2)\left(-\frac{3}{2}\right)\left(\frac{1}{4\sqrt{2}}\right) + \left(\frac{1}{\sqrt{8}}(6(4))\right) + \left(\frac{1}{2\sqrt{2}}\right)(6)(4) = -\frac{48}{8\sqrt{2}} + \frac{12}{\sqrt{2}} = \frac{6}{\sqrt{2}} = \frac{3*\sqrt{2}^2}{\sqrt{2}} = 3\sqrt{2}$$

Question 22: E
Because it's given that statement (*) is false, then it is not true that "every day next week, Fred will do at least one maths problem". Fred only needs to fail to do a maths problem on, say, Wednesday, for (*) to be false, even if he does maths problems every other day besides Wednesday.

Question 23: F

$$a^x b^{2x} c^{3x} = 2\ln\left(a^x b^{2x} c^{3x}\right) = \ln(2)\ln\left(a^x\right) + \ln\left(b^{2x}\right) + \ln\left(c^{3x}\right) = \ln(2)$$

$$x\ln\left(a^x\right) + \mathrm{x}\ \ln\left(b^2\right) + x\ln\left(c^3\right) = \ln(2)x\left(\ln\left(ab^2c^3\right)\right) = \ln(2)x = \frac{\ln(2)}{\ln\left(ab^2c^3\right)}$$

Question 24: D
The roots differ by 2.

By the quadratic formula, the roots must be:

$$x = \frac{11 \pm \sqrt{(-11^2) - 4(2)c}}{4}$$

Since the roots differ by 2,

$$\frac{11 + \sqrt{(-11^2) - 4(2)c}}{4} - \frac{11 - \sqrt{(-11^2) - 4(2)c}}{4} \frac{2\sqrt{(-11^2) - 4(2)c}}{4} = 2$$
$$\sqrt{121 - 8c} = 4121 - 8c = 16121 - 16 = 8c105 = 8cc = \frac{105}{8}$$

Question 25: C
The answer is option C. Using just the runners' initials for simplicity, and writing ">" to mean "beat", we are told that: F > G, H > L, L > G, R > G It immediately follows that H > G, so G must have come last. Among the other four, we only know that H > L.

There are several ways to work out the number of orders from here: • There are 4! = 24 ways to order the four runners.

In half of them H > L, in the other half, L > H. So there are 12 orders with H > L.

Question 26: E
$Let\, u = 2^x$

The equation becomes $u^2 - 8u + 15 = 0$
Solving the equation by factorizing, we get $(u-3)(u-5) = 0$
Which yields $u = 3\; and\, u = 5$
Using $u = 2^x,\; 2^x = 3\; and\; 2^x = 5$

$$x\, lg_{10}2 = lg_{10}3\; and\, x\, lg_{10}2 = lg_{10}5$$

Hence,

$$x = \frac{lg_{10}3}{lg_{10}2}$$

and

$$x = \frac{lg_{10}5}{lg_{10}2}$$

The sum of the roots is $\frac{lg_{10}5}{lg_{10}2} + \frac{lg_{10}3}{lg_{10}2} = \frac{lg_{10}15}{lg_{10}2}$.

Question 27: F
As the polynomial has degree 5, its derivative will have degree 4 so may have up to four real roots, and so our graph may have up to four stationary points. The graph of dx dy either tends to $+\infty$ as $x \to \pm\infty$ or it tends to $-\infty$ for both; either way, it must cross the x-axis an even number of times. (Touching the axis without crossing results in a point of inflection, which we are not considering in this question.)

Question 28: E
The answer is option E. Statement 1 subtracts a + b from both sides. Statement 2 can be written as $(a-b)^2 \geq 0$. This is always true.
Statement 3 can be false if c is negative, for example a = 2, b = 1, c = −1.

Question 29: C

This is a quartic graph and it will intersect the x-axis a minimum of two times, implying a minimum of two real roots. We differentiate the equation to find the turning points to see if the graph will cut the x-axis more times.

The derivative is:

$$4x^3 - 12x^2 + 8x = 0x\left(4x^2 - 12x + 8\right) = 04x\left(x^2 - 3x + 2\right) = 04x(x - 1)(x - 2) = 0$$

Coordinates of the turning points are thus:

$$(0, -10),\ (1, -9),\ (2, -10)$$

None of the turning points are between the positive and negative sides of the y axis, implying that there are only two real roots.

Question 30: D

The answer is option **D**.

The answer is $y = ax^b$

We can see this by taking logs of the equation and comparing it with the standard equation of a line,

$Y = mX + c$ Taking logs gives:

$\log y = \log a + b \log x$

Comparing with $Y = mX + c$, we can see that plotting $\log y$ on the Y-axis and $\log x$ on the X-axis would give us a straight line with a gradient of b and a Y-intercept of $\log a$.

Question 31: A

Integrate directly to get:

$$\left[\frac{(x-a)^3}{3}\right]_0^1 = a^2 - a + \left(\frac{1}{3}\right) \text{Complete the square to get:}$$

$$a^2 - a - \left(\frac{1}{2}\right)^2 + \left(\frac{1}{2}\right)^2 + \left(\frac{1}{3}\right)\left(a - \frac{1}{2}\right)^2 + \frac{1}{2}$$

The minimum value of the integral as a varies is $\frac{1}{12}$.

Question 32: E

The answer is option E.

The largest median will be if the middle number is as large as possible.

If the largest of the five numbers is u, then we will make the third number also u. Since the range is 20, the smallest number is $u - 20$.

To make the median u as large as possible, we want the second smallest number to be $u - 20$ too.

So the five numbers are $u - 20, u - 20, u, u, u$.

Their sum is $5u - 40 = 0$ as the mean is 0, so $u = 8$, which is the largest possible median.

Thus the answer is E.

Question 33: D

The answer is option D. For real distinct roots the discriminant condition gives:

$$(a-2)^2 > 4a(-2)a^2 + 4a + 4 > 0(a+2)^2 > 0$$

Which is true for all values of a except -2.

Question 34. C

The answer is option C.

$$P \propto \left(\frac{1}{Q^2}\right) P = \left(\frac{k}{Q^2}\right) \ for\ some\ k\ .$$

When Q increases by 40%, the new value is 1.4Q.

$$Q_{new} = 1.4Q$$

Which yields:

$$P_{new} = \frac{k}{(1.4Q)^2} = \frac{k}{1.96Q^2} = \frac{1}{1.96}P$$

We then notice:

$$\frac{1}{1.96} \approx \frac{1}{2}\ and\ \frac{1}{1.96} > \frac{1}{2}$$

Hence P_{new} is slightly over 50% of P and we can deduce that P has decreased a little under 50%.

Question 35: E

The answer is option E.

The probability that the first ball is red is $\frac{x}{x+y+z}$

The probability that the second ball is blue is $\frac{y}{x+y+z}$

The probability that the first ball is red and the second is blue is given by:

$$\frac{xy}{(x+y+z)^2}$$

Question 36: D
The answer is option D.

The first, second, and fourth terms of the GP are 4, 4r, and $4r^3$ respectively.

Since the terms are also in arithmetic series, the common difference is constant.

$$4r = 4 = 4r^3 - 4r$$

Factorizing this, we get:

$$4(r - 1) = 4r(r^2 - 1)$$

Dividing by 4(r-1) since r =/= 1

$1 = r(r + 1)$ yielding $r^2 + r - 1 = 0$

$$Solving: r = \frac{1}{2}(\sqrt{5} - 1)$$

Sum to infinity:

$$\frac{a}{1 - r} = \frac{8}{3 - \sqrt{5}}$$

Rationalizing this, we get: $2(3 + \sqrt{5})$

Section 2

What is understood by 'the Conventional Wisdom'? Discuss an example of an idea which qualifies as conventional wisdom.

An example of an idea that qualifies as conventional wisdom is the idea that individuals should choose college majors that correspond to 'useful' professions in society such as law or medicine, and that students who do not choose such majors effectively have shut themselves out of a future and will end up wasting their parents' money.

The reason that this constitutes a conventional wisdom is that it is an idea based on an assessment of stability and an assessment of the requirements of the real world, which is difficult and forbidding to engage with: The journey of an individual into the world of jobs and careers is a non-linear one, and upon first sight, majors such as law and medicine provide a pathway into the job market that is secure and free from uncertainty, even if it is true that such a pathway does not correspond to the interests of a particular student.

The dangers of promoting only a small class of academic fields to the exclusion of others has its roots in practicality, but may not serve the populace well whether it is with respect to personal satisfaction or professional success: To choose an academic major on the basis of what someone else has said is to effectively subjugate one's own decision to the judgment of others in order to make a call about what is best for oneself, and therefore to not pursue one's true passion in the pursuit of wealth from a path that has already been well-trodden and understood. While it is true that not every academic major trains individuals to directly pursue a position within the job market directly as with law and medicine, education is not simply for the purpose of obtaining employment, but is also for the purpose of ensuring that one can achieve fulfilment in one's life, for the purpose of shaping the events of the world that have yet to occur and to understand their trends rather than to simply go along with the trends of society and follow along with the current.

To claim that a major is useless for a particular purpose is to underestimate the value of the individual who stands behind the major rather than to definitively state that the major itself is useless, for each person who goes into the job market must make use of their skills, talents, and inclinations in order to stand and distinguish themselves in the job market: To designate a particular individual as being unemployable just because they have chosen a specific field of study is to place more credit on the past history of the individuals who have gone through a specific path rather than to give strength to the resolute determination of an individual who is driven to succeed in his or her chosen profession or pathway in life: Not everyone who is an investment banker or a management consultant has a major in economics, and indeed, many people come to a disparate variety of professions from majors in art history, commerce, and English literature with the skills, know-how, and preparation that is necessary to succeed in a career in virtue of their preparation and their desire for a career that is not at odds with their academic choices.

What the conventional wisdom proposes is a world where we remain in our lanes, pursuing what exists within the created world rather than rising up to create our own through the skills and experiences, the toolkits that we have developed in line with our specialization and interest to the exclusion of other worlds and pursuits, yet we live today in a world that is increasingly interconnected, where lines of knowledge seamlessly weave together, information is shared across worlds, and people step in and out of blurred boundaries across multiple streams of shared consciousness: This is a new world in which our contribution consists not simply in pushing into the workplace what we have learned in the academy, but also in piecing together what we see in order to make sense of a chaotic world.

While law and medicine are dreams that parents worldwide seek for their children, they are not the only dreams – For meaning is made from the pursuit of each path through a relentless process of action, a chase after success, and the cognition of a person as they progress through the world: Through choosing the academic and professional path that one truly desires, rather than a path that someone else has decided upon simply because it is 'stable', 'reputable', or 'desirable': Within the conscious pursuit of that path, even if it is not clear and the uncertainty is palpable, lies the answer to the questions that are posed within one's desire for self-fulfilment.

END OF PAPER

2016

Section 1A

Question 1: D

I leave London at 17:30 GMT on August 19th, which is at this time 20th August in Auckland, at 5:30. I travel to Auckland and arrive at 6:15 on August 21st, which means that my entire journey must have taken 24 hours and 45 minutes. Accounting for my 1 hour stop in Los Angeles and 1.5 hour stop in Hawaii, it must be the case that I am due to spend 22 hours and 15 minutes in the air.

Question 2: C

Of the options given, Kingda Ka is the only option that meets all the constraints:
Minimum height: 122cm
Speed: 128mph
Flight cost * 2: $56*2 = $102
Average queue: 39 minutes

Question 3: C

C is the only net that could be folded to make the above cube. A is not possible as the F will be in the wrong orientation. B is impossible because the E would be in the wrong position. D would not work because the E would be improperly aligned. F is eliminated from consideration because the B is in the wrong position.

Question 4: C

Robert is on duty for three nights and is on duty on Friday night. Tom is on duty for four nights, and thus in order to satisfy the regulation of no more than two consecutive nights, he must be working on Monday, Tuesday, Thursday, and Friday. This leaves Monday, Tuesday, Wednesday, and Thursday as possible slots for Sheila. Sheila and Rob must work on Wednesday, as a minimum of two guards must guard the premises each day and Tom's slots are already occupied. Thus, Rob must also work on Tuesday, as if he worked on Thursday, this would violate the regulation. As a result, Sheila's slots must be Monday, Wednesday, and Thursday.

Question 5: C

The cheapest possible option is to buy a concession family ticket for grandpa, dad, me, and for my brother who is two years older than me, although unfortunately, since we have decided to go on Saturday afternoon, we are subjected to peak hour prices. Buying a concession ticket is possible because grandpa is a senior citizen, my father is unemployed, and we are both children. This costs £4.80. Mother will pay for a £3.80 ticket, and grandma will pay £0.70 to observe. This yields a total cost of £9.30, and the answer is C.

Question 6: B
B does not correspond to any of the tiles on the floor, even if it is rotated. Hence, the answer is B.

Question 7: C
The fastest route is to travel from Essover to Yewton (20km), Yewton to Arford (18km), Arford to Teechester (25km), and then Teechester to Essover (23km). The sum of these distances is 86km.

Question 8: A
We subtract the total number of fatal injuries in non-built up areas (237) from the number of fatal or serious injuries (1724) to get a total of 1537 serious but not fatal injuries.

Question 9: E
From the directional perspective indicated, the leftmost column should be one block high, the second column from the left should be two blocks high, the middle column is 3 units high, and the last column should be two blocks high. Hence, the answer is E.

Question 10: B
The maximum score is 160, and we scored 138. If we answered 9 questions incorrectly and more frequently in Round 1 than any of the other rounds, the sole possibility for the distribution of our mistakes across rounds are:
Round 1: 5 mistakes (-5 points)
Round 2: 1 mistake (-2 points)
Round 3: 3 mistakes (-15 points)
If any other combination is tried, the numbers will not sum together such that we obtain 138 points total.

Question 11: D
I will buy a pack of 6 rolls at £20.94, as this is the most economical option.

Question 12: A
W fits onto the right side of Y, to and X fits into the gap on the right. Z can then be rotated 180 degrees and placed in the resulting gap to create a carpet of width of 3 and length of 6. V does not participate in this arrangement.

Question 13: A

The area of the entire flysheet is 24cm*18cm = 432cm^2. The text must occupy half of the total area, and thus it occupies 216cm^2.

Let us denote the margin width as x.

We know that the area of the text is 216cm^2, and so to solve for x, we get:

$$(24 - 2x)*(18 - 2x) = 216432 - 48x - 36x + 4x^2 - 216 = 0216 - 84x + 4x^2 = 0$$

Which yields $x = 3\ and\ x = 18\ as\ possible\ roots.\ We\ select\ x = 3.$

The answer is A, 3.00cm.

Question 14: A

The cheapest option for them is to sit in the Balcony section between Monday to Friday, as the Dress Circle seats are more expensive even with the Dress circle group booking discount and the box seats are still more expensive than the Balcony section seats.

Question 15: E

Pip: £54 x 28 = £1512
Eve: £56 x 27 = £1512
Nan: £72 x 21 = £1512
Bob: £63 x 24 = £1512
Viv: £68 x 22 = £1496

Viv is the only one with £1496. Hence, E is the answer.

Question 16: D

The question specifies that charges for coffee and tea are whole number amounts. In other words, for:

$$5c + s = 1214t + s = 82$$

s should be a number such that 121 – s and 82 – s are divisible by 5 and 4 into whole numbers respectively. The minimum number that satisfies this condition is s = 6.

Question 17: A

The total area covered by the mushrooms is 100m * 100m = 10000m^2, and we observe that the area expands such that A = (2n)^2, where n = 1,2,3….

The area covered by the mushrooms equals to that of the field when 10000 = (2n)^2, 100 = (2n), n =50. On Wednesday, two days would have passed since the beginning, and thus 48 days remain until the field is completely covered.

Question 18: D

A is represented by the second graph, B is represented by the fourth, C is represented by the first, and E is represented by the third. Only D is without representation.

Question 19: C
We know that the quantity that Anna can obtain is the budget that she has allocated, divided by the respective prices. Dividing £3.00 by a weekday price of 20p, we have a quantity of 15. If we divide £3.00 by the Saturday morning price, 25p, we have a quantity of 12 (3 lower than the quantity obtained under weekday price), and if we divide £3.00 by the Saturday closing price, 15p, we get a quantity of 20 (5 larger than the quantity obtained under the weekday price). The answer is C.

Question 20: A
The only cars that will satisfy the 4-5 door and 1.4 litre and above capacity constraint are the Rover 820 4 door, the Renault Laguna 5 door, Rover 825 4 door, and Ford Sierra 4 door.

Because he cares about lowest depreciation per mile, he will buy the Ford Sierra ($2000/30,000 miles) = $0.067 per mile.

Question 21: D
D is not true. The sum total of individuals both male and female in the 80-84 age group is 1332700. The sum total of individuals both male and female in the 85+ age group is 1274300, which is less than 1332700. Therefore, not taking children below the age of 1 into consideration, the least numerous age group is the 85+ age group.

Question 22: B
Imagine that the planets start off in alignment at time t = 0. One approach is to write expressions for the total full and partial revolutions that each planet will undergo in time t:

p1 = t/20
p2 = t/45
p3 = t/120

Planets line up again when p2-p1, p3-p2, and p3-p1 are all integer numbers of revolutions, i.e. all planets have moved the same amount round their orbits, give or take a whole number of revolutions.

p2-p1 = t/20 - t/45 = 25t/(20*45) = N (i.e. some integer) for alignment

=> t = N*(20*45)/25 = N*36

so Othello and Hamlet align every 36, 72, 108 days etc.

Similarly, you get a sequence of values for p3-p2 and for p3-p1. The smallest value of t these all have in common is your answer.

Section 1B

Question 23: D

The first three terms of a geometric progression are equivalent to the first, fifth, and sixth terms of an arithmetic progression. In other words,

$$a = aar = a + 4dar^2 = a + 5d$$

Rearranging the second equation to get r in terms of a and d, we get:

$$ar = a + 4dr = 1 + \frac{4d}{a}$$

Using the value of r that we obtained in the rearrangement of the previous equation in the third equation, we get:

$$a\left(\frac{a+4d}{a}\right)\left(\frac{a+4d}{a}\right) = a + 5d\frac{a^2 + 8ad + 16d^2}{a} = a + 5da^2 + 8ad + 16d^2 = a^2 + 5ad$$

$$3ad + 16d^2 = 0d(3a + 16d) = 03a + 16d = 03a = -16da = -\frac{16d}{3}$$

Substituting a into the equation that we had earlier for r, we get

$$r = 1 + \frac{4d}{\left(-\frac{16}{3}\right)d}r = 1 - \frac{3}{4}r = \frac{1}{4}$$

Question 24: D

The equation $2x^2 + 9x - k = 0$ has two roots, where one is more than the other by 4.

$$2x^2 + 9x - k = 0x^2 + \frac{9}{2}x - \frac{k}{2} = 0x = \frac{-\left(\frac{9}{2}\right) \pm \sqrt{\left(\frac{9}{2}\right)^2 - 4(1)\left(-\frac{k}{2}\right)}}{2(1)}$$

We know that one root is greater than the other by 4, and hence:

$$\frac{-\left(\frac{9}{2}\right) + \sqrt{\left(\frac{9}{2}\right)^2 - 4(1)\left(-\frac{k}{2}\right)} - \left(-\left(\frac{9}{2}\right) - \sqrt{\left(\frac{9}{2}\right)^2 - 4(1)\left(-\frac{k}{2}\right)}\right)}{2(1)} = 4$$

$$-\left(\frac{9}{2}\right) + \sqrt{\left(\frac{9}{2}\right)^2 - 4(1)\left(-\frac{k}{2}\right)} + \left(\frac{9}{2}\right) + \sqrt{\left(\frac{9}{2}\right)^2 - 4(1)\left(-\frac{k}{2}\right)} = 8$$

$$2\sqrt{\left(\frac{9}{2}\right)^2 - 4(1)\left(-\frac{k}{2}\right)} = 8\sqrt{\left(\frac{9}{2}\right)^2 - 4(1)\left(-\frac{k}{2}\right)} = 4\sqrt{\left(\frac{9}{2}\right)^2 + 2k} = 4$$

$$\left(\frac{9}{2}\right)^2 + 2k = 16\frac{81}{4} + 2k = \frac{64}{4}2k = -\frac{17}{4}k = -\frac{17}{8}$$

Question 25: D

The outer diameter of the roll is 11cm, or 110mm. The inner diameter is 5cm, or 50mm. The difference in diameter roughly attributable to the towels wrapping around the inner tube is 60mm.

Each towel is 250mm long, and there are 64 towels and thus 16000mm length of towel to wrap around the inner 5cm roll. Since the inner diameter is 5cm, the radius is 2.5cm, which is equivalent to 25mm, which means that the circumference is roughly $50\pi\,mm$, which is roughly 150mm. In other words, the 16000mm length of towels will wrap around the inner roll approximately 100 times in total.

The 100 wraps account for about 60mm diameter difference, and hence each individual towel accounts for approximately 0.5mm of the thickness of the roll.

Question 26: C

The initial value of the expression is:

$$\frac{abc(a+b+c)+\ 2d(3bcd)}{2d(a+b+c)}$$

Thus, the value of the expression where a, b, c, and d are increased by 20% each is.

$$\frac{(1.2a)(1.2b)(1.2c)(1.2a+1.2b+1.2c)+\ 2(1.2d)\left(3(1.2b)(1.2c)(1.2d)\right.}{2(1.2d)\left((1.2a)+(1.2b)+(1.2c)\right)}$$

$$=\frac{(1.2^4)abc(a+b+c)+1.2^4\left(2d(3bcd)\right)}{2(1.2^2)d(a+b+c)}=\frac{1.2^2(abc(a+b+c)+2d(3bcd))}{2d(a+b+c)}$$

1.2^2 is equivalent to 1.44.

Hence, we know that the percentage increase in the value of the expression is 44%.

Question 27: F

We are given that the perimeter of the sector is 12cm, and the length of an arc subtended by an angle θ radians is $r\theta$. We know thus that: $x + x + x\theta = 12cm$.

We know further that the area of the sector is defined by: $Area = \frac{1}{2}r^2\theta$.

Writing θ in terms of x:

$$x + x + x\theta = 12cm 2x + x\theta = 12 \frac{12 - 2x}{x} = \theta$$

$$Area = \frac{1}{2}r^2\theta Area = \frac{1}{2} * r^2 * \frac{12 - 2r}{r} Area = \frac{1}{2}r(12 - 2r) Area = 6r - r^2$$

For maximum area, we set $\frac{d(Area)}{dr} = 0$

$$\frac{d(Area)}{dr} = 6 - 2r = 0 r = 3$$

Substituting the value of r into $\frac{12 - 2r}{r} = \theta$, we get $\theta = 2$.

Using the Area formula and the values of θ and r that we obtained…

$$Maximum\ area = \frac{1}{2}r^2\theta Maximum\ area = \frac{1}{2}(3^2)(2) Maximum\ area = 9$$

Question 28: E

The equation of the gradient to the tangent to $y = 2x^3 - 9x^2 + 12x + p$ is $\frac{dy}{dx} = 6x^2 - 18x + 12$. To find the values of p, we can set $6x^2 - 18x + 12 = 0$ to find the points where the gradient to the tangent will be 0.

$$6x^2 - 18x + 12 = 0(3x - 6)(2x - 2) = \ 0x = 2\ x = 1$$

Substituting these x-values into the original equation, we find that the corresponding y coordinates for these points would be:

$$y = 5 + py = 4 + p$$

We are given that the graph only cuts the x-axis at one point. If the graph is to do so, it follows that the bottom turning point should not touch the graph. In order for the turning point to not touch the graph, p should be greater than 4.

Question 29: B

Using the properties of logarithms,

$$2 + \log_5 x^2 = \log_5(24 + 10x)$$

$$= \ \log_5 25 + \log_5 x^2 = \log_5(24 + 10x) = \ \log_5 25x^2 = \log_5(24 + 10x)$$

So we know that:

$$25x^2 = 24 + 10xx^2 - \frac{2}{5}x - \frac{24}{25} = 0$$

Solving for the value of x using the quadratic formula, we find that:

$$x = \frac{\left(\frac{2}{5}\right) \pm \sqrt{\left(\frac{4}{25}\right) - 4(1)\left(-\frac{24}{25}\right)}}{2} x = \frac{\left(\frac{2}{5}\right) \pm \frac{10}{5}}{2} x = \frac{6}{5}\ ,\ x = \ -\frac{2}{5}$$

$x = \frac{6}{5}$ is the positive value of x that we choose.

Question 30: B

Jez cycled the first 5km at 10km/h, and therefore must have taken ½ an hour to travel that distance. He travelled the remaining 10km at 30km/h, suggesting that he took 1/3 of an hour to travel 10km. His average speed is his total distance travelled divided by the total number of hours, which is 15km divided by (5/6) hours. His average speed is 18km/h.

Question 31: D

The equation of the curve is $y = x(x+a)(x-2a) = (x^2 + ax)(x - 2a)$
$= x^3 - 2ax^2 + ax^2 - 2a^2x = x^3 - ax^2 - 2a^2x$

The area of the curve enclosed within the curve and the lines x = -a and x = a, where a is a positive constant, is:

$$\int_{-a}^{a} x^3 - ax^2 - 2a^2x\,dx = \int_{0}^{a} x^3 - ax^2 - 2a^2x\,dx + \int_{-a}^{0} x^3 - ax^2 - 2a^2x\,dx$$

$$\int_{0}^{a} x^3 - ax^2 - 2a^2x\,dx = \left(\frac{a^4}{4}\right) - \frac{a^4}{3} - a^4 = -\frac{3a^4}{4} - \frac{4a^4}{12}$$

$$= -\frac{9a^4}{12} - \frac{4a^4}{12}$$

$$= -\frac{13a^4}{12}\int_{-a}^{0} x^3 - ax^2 - 2a^2x\,dx = \frac{a^4}{4} - \frac{a(-a^3)}{3} - \frac{2a^2(-a^2)}{2} = \frac{a^4}{4} + \frac{a^4}{3} - a^4$$

$$= -\frac{3a^4}{4} + \frac{a^4}{3} = -\frac{9a^4}{12} + \frac{4a^4}{12} = -\frac{5a^4}{12}$$

$$\int_{-a}^{a} x^3 - ax^2 - 2a^2x\,dx = \int_{0}^{a} x^3 - ax^2 - 2a^2x\,dx + \int_{-a}^{0} x^3 - ax^2 - 2a^2x\,dx$$

$$= -\frac{18a^4}{12} = -\frac{3}{2}a^4$$

The total area is $\frac{3}{2}a^4 units^2$.

Question 32: D

The x coordinates for points in the line will be the same. Distance will hence be:

$$Distance = \sqrt{((3x^2+2)-(5x-6))^2} Distance = \sqrt{((3x^2-5x+8))^2}$$

$$Distance = 3x^2 - 5x + 8$$

For minimal distance, we take the derivative of the distance formula and set that equal to 0, as follows:

$$\frac{d(Distance)}{dx} = 6x - 5 = 0x = \frac{5}{6}$$

For minimal distance, we substitute this value of x into the equation, yielding Distance $= \frac{71}{12}$.

Question 33: E

We seek the probability that at least one of the balls will be red. In other words, we seek:

$$P(at\ least\ one\ of\ the\ balls\ is\ red) = 1 - P(all\ the\ balls\ are\ blue).$$

$$P(all\ the\ balls\ are\ blue) = \frac{2n}{3n} * \frac{2n-1}{3n-1}$$

$$P(at\ least\ one\ of\ the\ balls\ is\ red) = 1 - P(all\ the\ balls\ are\ blue).$$

$$= 1 - \frac{4n-2}{3(3n-1)} = \frac{9n-3-4n+1}{3(3n-1)} = \frac{5n-1}{3(3n-1)}$$

Question 34: A

Statement 1 could be false if a and b are negative.

Statement 2 must be true regardless of the circumstances. If a is less than or equal to b, the expression with a will always be smaller than the expression with b.

Statement 3 need not be true. If c is initially negative and a is less than or equal to b and is negative, the statement will be false.

Question 35: E

We know that there is a maximum stationary point at x = 0 and a minimum stationary point in the 4th quadrant.

Taking the derivative of the curve $y = ax^3 + bx^2 + c$, we find that:

$$\frac{dy}{dx} = 3ax^2 + 2bx \frac{d^2y}{dx^2} = 6ax + 2b$$

In order that there be a maximum stationary point at x = 0, $\frac{d^2y}{dx^2} < 0$.

Hence, $6a(0) + 2b < 0$

Hence, $b < 0$.

In order that there be a minimum stationary point at a value in the fourth quadrant, $\frac{d^2y}{dx^2} > 0$.

$$6ax + 2b > 0 6ax > \; -2b$$

In the fourth quadrant, x > 0 and from the previous stationary point, we know that b < 0. The quantity (-2b) can be represented therefore as a positive real number, which we can denote as c.

6ax > c

$a > \frac{c}{6x} > 0$

Hence, a > 0.

E is the only option that presents a>0 and b<0.

Question 36: E

If bulb X is off or bulb Y is on, then bulb Z is on.

If bulb Z is off, then this is necessarily because bulb X is on or bulb Y is off.

Question 37: E

The mean mass of the group is initially 84kg.

Thus, the total mass of the group of 20 individuals must be 84*20 = 1680kg.

We are asked for an expression for the mean mass of N individuals who leave the group and cause the mean to become 81kg.

We know thus that:

$$\frac{1680 - N(\mu)}{20 - N} = 811680 - N(\mu) = 1620 - 81N60 = N(\mu) - 81N60 = N(\mu - 81)$$

$$\frac{60}{N} + 81 = \mu$$

Hence, the answer is E.

Section 2

How does cooperation between people take place when no one is explicitly in charge?

Cooperation between people takes place when individuals cooperate to meet their best interests and needs through means that are self-interested in motivation, but socially beneficial to the world. Indeed, the famous Adam Smith quote, "It is not from the **benevolence of the butcher**, the brewer, or the **baker** that we expect our dinner, but from their regard to their own interest." Indeed, even if nobody is explicitly in charge of coordinating cooperation, the millions of decisions that people make throughout the world to fulfill their own needs creates the grand pattern of cooperation that we see throughout the world, as each acts according to his own need, with the resources that are available at hand.

Take the example of the shirt that is provided in this essay, for example. As the author notes, "the shirt I bought, although a simple item by the standards of modern technology, represents a triumph of international cooperation. The cotton was grown in India, from seeds developed in the United States; the artificial fibre in the thread comes from Portugal and the material in the dyes from at least six other countries; the collar linings come from Brazil, and the machinery for the weaving, cutting, and sewing from Germany; the shirt itself was made up in Malaysia."

Why is it that these things occur, and how is it that individuals from such disparate geographical regions cooperate with one another over such a large scale in such a way? It is certainly not the case that a central entity is coordinating the production of these individual components - Rather, cooperation across these jurisdictions is the product of individuals making use of the idiosyncrasies of each location and translating them into products that it is most suited to make. People cooperate with one another when they can achieve specific goals, and one specific goal that most individuals have in mind is making a living: Each land has its own resources, capacities, and people of skill such that the land develops and people seek to make something of themselves.

Naturally, in a shared world, there are different kinds of lands, and different kinds of individuals, each with his own talents varying across the entire scope of human ability, as well as a differential ability to make use of those talents depending on the endowment of resources and market environment within a country. India, for example, has a climate that is especially suited for growing cotton, while Germany has a great industrial production operation that has been built on the back of engineering expertise, each of which has made these countries suited to performing the specific tasks that lead into the production of the shirt. If instead Germany had tried to grow cotton and India had tried to create machinery and neither of these countries had the expertise to do so, they would waste more time on these tasks rather than creating the products beforehand and exchanging them for the resources of other entities, as they would not have the people or the capacity necessary to participate in the market.

To the extent that human needs are common and are shared amongst individuals, it is possible for businesses to plan out the ways in which their labour will be distributed and sold with reasonable certainty of the ways in which these needs may be fulfilled, even if it is true that these individuals do not directly or explicitly cooperate with one another: To get the cheapest, most high quality material is the prerogative of the shirt manufacturer, who buys from the region where the product can be produced at both quality and cost in accordance with the specialization that arises from the development of a nation's comparative advantage, while providing the cheapest, highest quality material is the prerogative of the seller who has as his interest outcompeting other parties who would like to sell within the same market to other parties who are keen on converting the material into the infinitude of products that make up the sheer complexity that is our modern age market... None of this coordination requires the conscious will of a single individual, and would certainly exceed the capacity of that individual if he were to attempt to replicate it.

END OF PAPER

2017

Section 1A

Question 1: D

The height of the picture is 40cm, and the width of the frame it is contained in is 2cm. The mount is 6cm wide at the top, and 9cm wide at the bottom. The overall height of the picture is obtained by summing all the heights: 40cm + (2cm + 2cm) (height of the frame) + mount width (6cm + 9cm) = 59cm.

Question 2: D

Rhine is the answer. Look at the pairings across rows and columns – Rhine is the only team to a positive score for every single row and column pairing in which it appears.

Question 3: A

The total area that Rosie needs to mow is Area to mow = (10m * 25m) – (2m * 3m) = 244m^2. With her old mower, she would have taken 244 minutes to mow the lawn at 1m^2 per minute. With her new mower, she takes 122 minutes to mow the lawn at 2m^2 per minute. The time savings that she sustains is 244 – 122 = 122 minutes.

Question 4: C

My friend has a minimum of 1 coupon of each denomination, which make up a total of (7p + 12p + 19p) = 38p value. To find the number of 19p coupons that he has, we can consider the case in which he has the fewest possible number of 19p coupons. Suppose that my friend had four 12p coupons in addition to the three he already has – His total coupon value would be 86p. In order that his total coupon value be 1 pound in total, it must be the case that two of the coupons have an individual value of 7p each. In other words, he needs three 19p coupons.

Question 5: C

We need to find a difference in points that corresponds to 214 points for an 18cm record improvement. Looking at the table, we can see that the difference between 1.74m and 1.56m is 214 points. Hence, C is the answer.

Question 6: D

We take the difference between car registrations for 2007 and 2012 for each month of the year.

January: 37.1
February: 10.1
March: 77.4
April: 29.9
May: 25.8
June: 35.1
July: 33.6
August: 18.7
September: 60.1
October: 18.3
November: 10.2
December: 16.7

The only graph provided that represents these totals accurately in terms of relative size is graph D.

Question 7: A

To paint the wall and ceilings completely with the exception of the 10m^2 combined surface area occupied by the windows and door, I will need 2*(3m * 8m) + 2*(3m*4m) + (8m*4m) – 10m^2 = 94m^2 of paint. Each tin contains 8 litres of paint, and thus will allow me to cover (8 * 12m^2) = 96 m^2 of surface area.

Hence, one tin of paint is sufficient.

Question 8: E

The friends' preferences are shown:

1st choices: Portugal Portugal Greece Majorca Tenerife
2nd choices: France France France France Tenerife
3rd choices: France Tenerife Tenerife Greece Majorca

Greece corresponds to one first choice and one third choice, and thus to 3 + 1 = 4 points. Portugal corresponds to 3 + 3 = 6 points. Majorca corresponds to 3 + 1 points. Tenerife corresponds to one first choice, one second choice, and two third choices, and thus to 3 + 2 + 1 + 1 points = 7 points. Hence, the friends go to Tenerife.

Question 9: B

The vase is curved and concave, and thus the increase in height must be accelerating from the start, and begin to decelerate as the vase becomes wider. The only graph that satisfies this condition is graph B.

Question 10: B

The information given is that 6.0kg of the mixture contains ¼ of X and ¾ of Y, which means that there is 1.5kg of X and 4.5kg of Y in the solution. The technician wants to create a solution with the proportions 40% of X and 60% of Y, and hence the addition of a mass of X must result in a chemical 40% composed of X.

We can find this mass by solving the equation:

$$\frac{1.5+x}{6+x} = 0.4$$

Solving for x, we find that $x = 1.5$ kg.

Question 11: D

Observe that the total mass of the package is 300g. Using the typical values per 100g, we can see that 300g of the material will contain 4.8g of sugars. Since each oatcake contains 0.3g of sugars, we can conclude that there is a total of (4.8/0.4) = 24 oatcakes.

Question 12: B

We are given the totals for various variables, and can fill in the remainder of the information from there. The total number of boys and girls is 900, and thus the number of boys is (900 – 460) = 440. Using this, we can solve for the number of people who take the bus, which is (440-80-130-200) = 30. The total number of walkers is 410, and 200 boys walk. Therefore, 210 girls must walk. Hence, the number of girls who commute by car must be (460 – 50 – 80 – 210) = 120.

Thus, the total number of people who commute by car is 120 + 80 = 200
The total number of people who commute by walking is 410
The total number of people who commute by biking is 210
The total number of people who commute by bus is 50 + 30 = 80

The only graph that accurately represents this information is graph B.

Question 13: C

To score 52 points, the Blues must have scored 8 majors and 4 minors, while to score 77 points, the Reds must have scored 7 majors and 14 minors. Hence, the total number of majors is 8 + 7 = 15.

Question 14: D

You can calculate three values. If you set up equations to represent each word, letting f,g,h,i represent each value:

T + E + A +R = f
R + I + T + E = g
T + R + E +E = h
R + A + T = i

You will see that E, A, I are the only equations that do not appear in all the equations and thus can be expressed in terms of the other variables. Hence, the answer is three variables.

Question 15: D

D is not possible to make, as the long side of the L would have to be on the opposite side for shape D to be created and would require a backwards L shape: The other shapes can all be made.

Question 16: C

If there were only three boats, they would respectively leave the island at 9:05am, 9:25am, and 9:45am, arrive at 9:45am, 10:05am and 10:25am, then take 40 minutes to return at 10:30 for departure at 10:35am, 10:50am for departure at 10:55am, and 11:10am for departure at 11:15am. However, the ferries begin leaving the island beginning from 9:15am, and if there were only three boats, it would be impossible to account for both the 9:15am and 9:35am departure time. Hence, a minimum of five boats is necessary.

Question 17: A

We can solve this by creating simultaneous equations to represent the final total of legs, horns, and tails that we will obtain from the capture of all three types of creatures.

Let A, O, and U represent the quantity caught of Arps, Orps, and Urps respectively.

3A + 2O = 99 horns
A + U = 33 tails
6A + 4O + 3U = 222 legs

Using the first equation, $A = \frac{99 - 2O}{3}$

Using the third equation,

$$6\left(99 - \frac{2O}{3}\right) + 4O + 3U = 222$$

$$198 - 4O + 4O + 3U = 222$$

$$198 + 3U = 222$$

$$3U = 24$$
$$U = 8$$

Hence, the answer is A.

Question 18: C

Suppose that I consume aloe vera juice at my regular dosage per day. I will consume the entire 600ml bottle in (600ml/dosage per day) days.

If I consume aloe vera juice at the slowed rate after two regular doses and realising that the supplier has sold out, the entire bottle will take 6 days longer to consume. In other words, accounting for the first two dosages in each bottle, (2*dosage + ((3/4)*dosage*(n+6)) = (2*dosage + dosage*(n)).

Solving for the number of days that I would take to finish the entire bottle with a regular dosage beyond the first two dosages, I set ((3/4)*dosage*(n+6)) = (2*dosage + dosage*(n)), which yields a value of n = 18, and showcases that in a regular scenario with a regular dosage, I would finish the entire bottle in 20 days.

Thus, my normal daily dose of aloe vera juice must be 600ml/20 days = 30 ml per day. The answer is C.

Question 19: B

The amount that Joan spends on cat food each day (2 sachets of wet food, 25g of dry food) when she buys from the local pet store is ((£12.00 for 12 sachets)/6) + ((£4.00 for 400 grams)/400*25 grams) = £2.25.

The amount that Joan spends on cat food each day when she buys online is (((£62.40 for four boxes)/4)/24 sachets)*2) + ((£16 for two kilograms)/2000*25) = £1.50.

The cost savings is £2.25 - £1.50 = £0.75 = 75p.

Question 20: D

At 8:00 to 20:00 local time, provided that London is used as the reference point, the siblings' availabilities are:

Nathan: 08:00 – 20:00
Mark: 00:00 – 12:00
Ben: 09:00 – 21:00
Isabel: 10:00 – 22:00

Between 00:00 to 09:00 only Nathan and Mark will be available. Beyond 20:00, only Ben and Isabel will be available. Hence, at least three siblings will be available simultaneously throughout the course of (24 – (9 + 4)) = 11 hours.

Section 1B

Question 21: D

The first price reduction brings the shirt to 70% of the original price. The second price reduction brings the shirt to 70%*80% = 56% of the original price. The price reduction relative to the original price is therefore 44% and the answer is D.

Question 22: E

$$\int_1^2 \left(3x + \frac{1}{x}\right)(3x + \frac{1}{x})\ \mathrm{dx}$$

$$= \int_1^2 \left(9x^2 + 6 + \frac{1}{x^2}\right) dx$$

$$= \left(9(2)^2 + 6 + \frac{1}{(2)^2}\right) - \left(9(1)^2 + 6 + \frac{1}{(1)^2}\right) = 35.5 - 8 = 27.5$$

Question 23: E

The inequality is:

$x - \frac{3}{2} > \frac{1}{x}\frac{2x-3}{2} > \frac{1}{x}2x^2 - 3x > 2x^2 - \frac{3}{2}x - 2 > 0$By the quadratic formula, the roots are $-\frac{1}{2}\ and\, 2$. Substituting the values into the equation, we find that the range of values is $-\frac{1}{2} < x < 0$, and $x > 2$.

Question 24: D

The ratio of those who study both languages to those who do not is 5:3. Therefore, the number of people who study no languages is 6. The intersection between French and German is 10 individuals, and therefore there must be 32 individuals who do not lie in the intersection and study only German and thus 75 – 32 – 6 = 37 individuals total who study French.

The probability that one pupil studies ONLY German is 32/75.

Question 25: A

$$y = 3 + 2\left(\frac{x}{4} - 1\right)^2 y - 3 = 2\left(\frac{x}{4} - 1\right)^2 \frac{y-3}{2} = \left(\frac{x}{4} - 1\right)^2 \pm\sqrt{\frac{y-3}{2}} = \left(\frac{x}{4} - 1\right)$$

$$1 \pm \sqrt{\frac{y-3}{2}} = \frac{x}{4}x = 4(1 \pm \sqrt{\frac{y-3}{2}})$$

Question 26: A

By the remainder theorem, dividing f(x) by (x+1) is analogous to evaluating f(-1), and dividing by (x-1) is analogous to evaluating f(1).

Hence,

$$f(-1) = 12 = 2(-1) + p(1) + q(-1) + 6$$
$$p - q = 8$$
$$p = 8 - q$$
$$f(1) = -6 = 2(1) + p(1) + q(1) + 6$$
$$p + q = -14(8 - q) + q = -14q = -11p = -3$$

Identically, dividing f(x) by (2x-1) to find the remainder is analogous to evaluating f(1/2), which we can now do since we know the values of p and q.

Hence, the remainder when f(x) is divided by (2x-1) is:

$$f\left(\frac{1}{2}\right) = 2\left(\frac{1}{8}\right) - 3\left(\frac{1}{4}\right) - 11\left(\frac{1}{2}\right) + 6 = 0$$

Question 27: F

The probability that the balls are not the same colour is 1 – the probability that the balls are the same colour. The probability that the balls are the same colour is the probability that a red ball is chosen and a red ball is chosen again, added to the probability that a green ball is chosen and a green ball is chosen again.

The probability that two balls chosen are the same colour is:

$$P(Same\ color) = \frac{n*(n+1)}{(3n)*(3n+1)} + \frac{(2n)*(2n+1)}{(3n)*(3n+1)} = \frac{1}{3}*\frac{n+1}{3n+1} + \frac{2}{3}*\frac{2n+1}{3n+1}$$

The probability that two balls chosen are not the same colour is therefore:

$$P(Different\ color) = 1 - \left(\frac{1}{3}*\frac{n+1}{3n+1} + \frac{2}{3}*\frac{2n+1}{3n+1}\right)$$

$$P(Different\ color) = \frac{9n+3-n-1-4n-2}{9n+3} = \frac{4n}{3(3n+1)}$$

Hence, the answer is F.

Question 28: E

We observe that the graph cuts the positive x-axis at x=3.

The gradient of the tangent is given by $\frac{dy}{dx} = 3x^2 - 7$

And thus the gradient of the tangent at x=3 is $\frac{dy}{dx} = 3(3)^2 - 7 = 20$.

The equation of the tangent is thus:

$$\frac{y-0}{x-3} = 20y = 20x - 60$$

Hence, the answer is E.

Question 29: A

$$2x - y = py = 2x - p3x^2 - x(2x - p) = 43x^2 - 2x^2 + px = 4x^2 + px = 4$$
$$x^2 + px - 4 = 0$$

There are two distinct and real solutions for x, and thus by the quadratic formula,

$b^2 - 4ac > 0p^2 - 4(1)(-4) > 0p^2 > -16$
Since every square number is greater than 0, therefore the answer is A.

Question 30: B

There are several unsurveyed individuals within the population. Looking at the totals, we can see that there are 94-(34+12+29) = 19 unsurveyed individuals in the population of category P employees, whereas there is a total of 86 – (30+21+27) = 9 unsurveyed employees in category Q.

The probability that an employee who commutes by car is in category Q would be lowest if all the unsurveyed category P employees commuted by car, and all the unsurveyed category Q employees commuted another way.

The probability that an employee commuting by car is in category Q, in that scenario, would be $\frac{21}{21+31} = \frac{21}{52}$. The answer is D.

Question 31: C

Since S and T are geometric progressions, and they have the same second term and sum to infinity, we can denote the first terms of S and T respectively as a_{1T} and a_{1S}, and the second terms of S and T as $T_{2S} = a_{1S}r_{1S}$ and $T_{2T} = a_{1T}r_{1T}$ and that both $T_{2S} = T_{2T} = 6$, but that $T_{1S} > T_{1T}$.

We know further that the sum to infinity of a geometric series, $S_\infty = \frac{a}{1-r}$ and that for S and T, , $S_\infty = \frac{a_{1T}}{1-r} = \frac{a_{1S}}{1-r} = 25.$

We know thus that $a_{1S} = 25 - 25r.$

Using $T_{2S} = a_{1S}r\ = 6$, we get:

$$(25 - 25r)r\ = 6(25r - 25r) = 625r - 25r + 6 = 0r^2 - r + \left(\frac{6}{25}\right) = 0$$

By the quadratic formula,

$$r = \frac{1 \pm \sqrt{1 - 4(1)\left(\frac{6}{25}\right)}}{2} r = \frac{1 \pm \frac{1}{5}}{2}$$

$$r = 0.6$$
$$r = 0.4$$

Are two possible values for r.

Now, there are two possible values for r, and we know that $T_{2S} = a_{1S}r_{1S}$ and $T_{2T} = a_{1T}r_{1T}$, and that $T_{1S} > T_{1T}$. Substituting the value of r_{1S} into the equation, $a_{1S} = \frac{6}{0.4} = 15$, $a_{1T} = \frac{6}{0.6} = 10.$

The question asks for the fourth term of S, $T_{4S} = a_{1S}r_S^3 = 15*(0.4)^3 = \frac{24}{25}$

Question 32: C

If 5 is added on to every term, then the sum of the series will have 5n added on to it, where n is the number of terms in the series, because each term is larger by 5 and there are still n terms in total in the new series, hence the answer is C.

Question 33: A

To find the value of x, we can simply perform a substitution after performing a multiplication, by observing that the equation:

$$5^{2x+1} + 5^x - 4 = 0$$

May be written as:

$5(5^{2x}) + 5^x - 4 = 0$

Here, we can perform a substitution and solve the equation as a quadratic, such that:

$u = 5^x$

The equation then becomes:

$$5u^2 + u - 4 = 0$$

Which factors to:

$$(5u - 4)(u + 1)$$

Which gives us roots $5u = 4$ and $u = -1$. Here, we re-use our substitution, $u = 5^x$, and, observing that this exponential function can never be negative for any value of x, eliminate u = -1 as a root.

We thus use the root 5u = 4.

$$5u = 45(5^x) = 45^{x+1} = 4(x+1)\ln(5) = \ln(4)(x+1) = \frac{\ln(4)}{\ln(5)}$$

Which, by change of base to base 2, becomes:

$$(x+1) = \frac{lg_2 4}{lg_2 5} * \frac{lg_e 2}{lg_e 2}(x+1) = \frac{2}{lg_2 5} x = \frac{2}{lg_2 5} - 1$$

Question 34: A

The intersections of the curves are located at x=1 and x=2.

The area between the two curves is:

$$\int_1^2 (-4 + 6x - 2x^2)dx = \frac{1}{3}$$

Question 35: D

- y = f(x)+2 need not necessarily intersect the x-axis as the graph has been translated upwards and may not even touch it.

  y = f(x) – 2 need not necessarily intersect the x-axis as the graph has been translated downwards and may not even touch it.
- y = f(x+2) necessarily intersects the x-axis at the same points, as the graph has simply been shifted leftward by two units.

  y = 2f(x) necessarily intersects the x-axis twice, as the graph has simply undergone a vertical stretch that does not affect whether the axes are cut by the graph or not.
- y = 2-f(x) need not necessarily intersect the x-axis as the graph has been shifted upward by two units.
- y = f(-2x) necessarily intersects the x-axis twice, as the graph has simply undergone a horizontal compression and this does not change the fact that it cuts the x-axis twice.

There are three scenarios above wherein the graphs necessarily intersect the graph at 2 distinct points. The answer is D.

Question 36: D

R is most definitely false, as I saw a pig with wings without horns.

P cannot be determined, as the pig had no horns and therefore I cannot make a statement about causality here.

Q cannot be determined. The pig has wings and it breathes fire, but I can conclude nothing about whether breathing fire relates directly to the pig having wings.

Section 2

Evaluate two arguments for and two arguments against low interest rates.
To what extent does the case for low interest rates depend on other policy choices?

The passage provides a host of arguments against lowered interest rates, as well as arguments that support lowered interest rates within the context of a climate of austerity amongst central banks and national economies, during which governments have opted to undertake contractionary fiscal policy so as to mitigate the problems that arise from an overheating economy.

The article argues that a climate of low interest rates can be beneficial for the economy, particularly in a climate of contractionary fiscal policy. Contractionary fiscal policy results in a lowering of aggregate demand as a result of lowered government spending and low interest rates act as a form of monetary stimulus that serves to shift aggregate demand outward, mitigating the negative consequences of tight fiscal policy. Specifically, the effects of tight fiscal policy may result in a lowering of aggregate demand within the economy, but governments may wish to limit the extent to which spending decreases by decreasing the financial reward for saving, albeit accepting the spending stimulus thesis without question requires the assumption that capital that would have been saved under a higher interest rate regime would be spent instead, which is not necessarily going to be true.

Above and beyond considerations of the actual empirical effect of an interest rate decrease on the level of aggregate spending in the economy, it is desirable to think of lowering interest rates as a form of monetary stimulus that is easy to execute by a central bank without the necessity of breaking through legislative gridlock or obtaining bipartisan agreement - Indeed, apart from ease of use, another argument in favour of lowering interest rates close to zero is that low interest rates do not have an obvious cost, as the risk of inflation is low in large economies due to ample spare capacity, and central banks are not subjected to the political constraints that their companions in government face when making decisions about fiscal policy, during which they must minimally ensure bipartisan agreement across the political divide concerning taxation and public spending.

However, low interest rate policy is not always ideal under all circumstances. Central banks face a problem that is called the zero lower bound problem: Banks cannot lower rates beyond zero, and monetary policy has limited usefulness beyond this point. There are several immediate arguments against low interest rates. The first of these arguments is that low interest rates can have a distortionary effect on capital, specifically that low interest rates can cause individuals to allocate capital toward unproductive projects that do not maximally benefit society.

The second of these arguments is that low interest rates may incentivize destabilizing behaviour such as capital flight from countries with low interest rates to jurisdictions with higher ones, as investors seek to place their capital wherever they can obtain the highest possible rate of return, as well as induce banks to take the risk of borrowing debt with short maturities to capitalize on the low short-term rate to lend at higher-yield long term rates, and thereby make themselves susceptible to the vagaries of interest rate change.

Taking both the advantages and disadvantages of lowered interest rates into account, however, requires adequate consideration of the current policy mix, specifically the specific objectives of the country with respect to inflation and unemployment, which in turn will depend on the economic numbers for a particular quarter, election promises that were made in previous cycles, and the actions of external investors. If a country is, for example, trying to deal with inflation, it may try to increase its interest rate to encourage saving instead of spending within the economy, but correspondingly attract hot money from foreign destinations seeking the promise of higher returns.

It behoves governments to not simply consider the possibility of confounding effects, however, but also the possibility of supporting other objectives - A good example is mitigating the impact on aggregate demand from a tightening fiscal policy, an objective that governments can achieve by lowering the interest rate to help support the objective of increasing say, consumer spending so as to soften the impact from contractionary policies, albeit such a policy mix must also depend upon the mixture of stakeholders that a government must appeal to at any given point in time - It may be more attractive to countries to achieve their policy goals by increasing the interest rate than to cut government spending on a particular program so as to appeal to certain stakeholders just before an election.

END OF PAPER

ECAA Practice Papers

Already seen them all?
So, you've run out of past papers? Well that is where this book comes in. This book contains two unique mock papers; each compiled by Cambridge economics tutors at *UniAdmissions* and available nowhere else.

Having successfully gained a place on their course of choice, our tutors are intimately familiar with the ECAA and its associated admission procedures. So, the novel questions presented to you here are of the correct style and difficulty to continue your revision and stretch you to meet the demands of the ECAA.

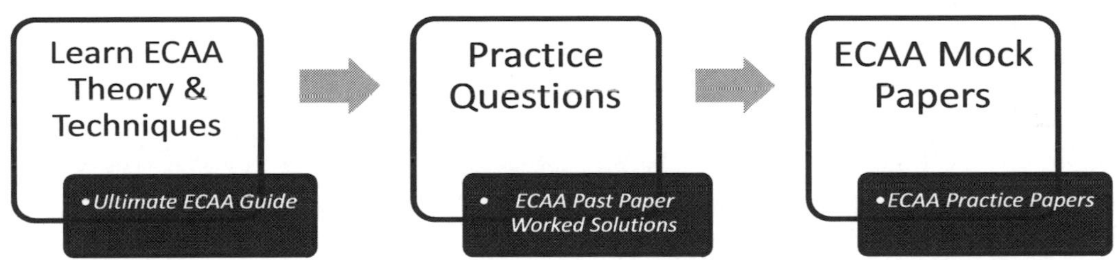

Start Early
It is much easier to prepare if you practice little and often. Start your preparation well in advance; ideally 10 weeks but at the latest within a month. This way you will have plenty of time to complete as many papers as you wish to feel comfortable and won't have to panic and cram just before the test, which is a much less effective and more stressful way to learn. In general, an early start will give you the opportunity to identify the complex issues and work at your own pace.

Positive Marking
There are no penalties for incorrect answers; you will gain one for each right answer and will not get one for each wrong or unanswered one. This provides you with the luxury that you can always guess should you absolutely be not able to figure out the right answer for a question or run behind time. Since each question provides you with 4 to 6 possible answers, you have a 16-25% chance of guessing correctly. Therefore, if you aren't sure (and are running short of time), then make an educated guess and move on. Before 'guessing' you should try to eliminate a couple of answers to increase your chances of getting the question correct. For example, if a question has 5 options and you manage to eliminate 2 options- your chances of getting the question increase from 20% to 33%!

Avoid losing easy marks on other questions because of poor exam technique. Similarly, if you have failed to finish the exam, take the last 10 seconds to guess the remaining questions to at least give yourself a chance of getting them right.

Prioritise

Some questions in sections can be long and complex – and given the intense time pressure you need to know your limits. It is essential that you don't get stuck with very difficult questions. If a question looks particularly long or complex, mark it for review and move on. You don't want to be caught 5 questions short at the end just because you took more than 3 minutes in answering a challenging multi-step question. If a question is taking too long, choose a sensible answer and move on. Remember that each question carries equal weighting and therefore, you should adjust your timing in accordingly. With practice and discipline, you can get very good at this and learn to maximise your efficiency.

Practice

This is the best way of familiarising yourself with the style of questions and the timing for this section. Although the exam will essentially only test GCSE level knowledge, you are unlikely to be familiar with the style of questions in all sections when you first encounter them. Therefore, you want to be comfortable at using this before you sit the test.

Practising questions will put you at ease and make you more comfortable with the exam. The more comfortable you are, the less you will panic on the test day and the more likely you are to score highly. Initially, work through the questions at your own pace, and spend time carefully reading the questions and looking at any additional data. When it becomes closer to the test, **make sure you practice the questions under exam conditions**.

Repeat Questions

When checking through answers, pay particular attention to questions you have got wrong. If there is a worked answer, look through that carefully until you feel confident that you understand the reasoning, and then repeat the question without help to check that you can do it. If only the answer is given, have another look at the question and try to work out why that answer is correct. This is the best way to learn from your mistakes, and means you are less likely to make similar mistakes when it comes to the test. The same applies for questions which you were unsure of and made an educated guess which was correct, even if you got it right. When working through this book, **make sure you highlight any questions you are unsure of**, this means you know to spend more time looking over them once marked.

Use the Options:

Some questions may try to overload you with information. When presented with large tables and data, it's essential you look at the answer options so you can focus your mind. This can allow you to reach the correct answer a lot more quickly. Consider the example below:

The table below shows the results of a study investigating antibiotic resistance in staphylococcus populations. A single staphylococcus bacterium is chosen at random from a similar population. Resistance to any one antibiotic is independent of resistance to others.

Calculate the probability that the bacterium selected will be resistant to all four drugs.

Antibiotic	Number of Bacteria tested	Number of Resistant Bacteria
Benzyl-penicillin	10^{11}	98
Chloramphenicol	10^{9}	1200
Metronidazole	10^{8}	256
Erythromycin	10^{5}	2

A. 1 in 10^{6}
B. 1 in 10^{12}
C. 1 in 10^{20}
D. 1 in 10^{25}
E. 1 in 10^{30}
F. 1 in 10^{35}

Looking at the options first makes it obvious that there is **no need to calculate exact values**- only in powers of 10. This makes your life a lot easier. If you hadn't noticed this, you might have spent well over 90 seconds trying to calculate the exact value when it wasn't even being asked for.

In other cases, you may actually be able to use the options to arrive at the solution quicker than if you had tried to solve the question as you normally would. Consider the example below:

A region is defined by the two inequalities: $x - y^2 > 1 \wedge xy > 1$. Which of the following points is in the defined region?

A. (10,3)
B. (10,2)
C. (-10,3)
D. (-10,2)
E. (-10,-3)

Whilst it's possible to solve this question both algebraically or graphically by manipulating the identities, by far **the quickest way is to actually use the options**. Note that options C, D and E violate the second inequality, narrowing down to answer to either A. or B. For A: $10 - 3^2 = 1$ and thus this point is on the boundary of the defined region and not actually in the region. Thus the answer is B (as 10-4 = 6 > 1.)

In general, it pays dividends to look at the options briefly and see if they can be help you arrive at the question more quickly. Get into this habit early – it may feel unnatural at first but it's guaranteed to save you time in the long run.

Manage your Time:
It is highly likely that you will be juggling your revision alongside your normal school studies. Whilst it is tempting to put your A-levels on the back burner falling behind in your school subjects is not a good idea, don't forget that to meet the conditions of your offer should you get one you will need at least one A*. So, time management is key!

Make sure you set aside a dedicated 90 minutes (and much more closer to the exam) to commit to your revision each day. The key here is not to sacrifice too many of your extracurricular activities, everybody needs some down time, but instead to be efficient. Take a look at our list of top tips for increasing revision efficiency below:

1. Create a comfortable work station: Declutter and stay tidy
2. Treat yourself to some nice stationery
3. See if music works for you à if not, find somewhere peaceful and quiet to work
4. Turn off your mobile or at least put it into silent mode and silence social media alerts
5. Keep the TV off and out of sight
6. Stay organised with to do lists and revision timetables – more importantly, stick to them!
7. Keep to your set study times and don't bite off more than you can chew
8. Study while you're commuting
9. Adopt a positive mental attitude
10. Get into a routine
11. Consider forming a study group to focus on the harder exam concepts
12. Plan rest and reward days into your timetable – these are excellent incentive for you to stay on track with your study plans!

Keep Fit & Eat Well:
'A car won't work if you fill it with the wrong fuel' - your body is exactly the same. You cannot hope to perform unless you remain fit and well. The best way to do this is not underestimate the importance of healthy eating. Beige, starchy foods will make you sluggish; instead start the day with a hearty breakfast like porridge. Aim for the recommended 'five a day' intake of fruit/veg and stock up on the oily fish or blueberries – the so called "super foods".

When hitting the books, it's essential to keep your brain hydrated. If you get dehydrated you'll find yourself lethargic and possibly developing a headache, neither of which will do any favours for your revision. Invest in a good water bottle that you know the total volume of and keep sipping throughout the day. Don't forget that the amount of water you should be aiming to drink varies depending on your mass, so calculate your own personal recommended intake as follows: 30 ml per kg per day.

It is well known that exercise boosts your wellbeing and instils a sense of discipline. All of which will reflect well in your revision. It's well worth devoting half an hour a day to some exercise, get your heart rate up, break a sweat, and get those endorphins flowing.

Sleep
It's no secret that when revising you need to keep well rested. Don't be tempted to stay up late revising as sleep actually plays an important part in consolidating long term memory. Instead aim for a minimum of 7 hours good sleep each night, in a dark room without any glow from electronic appliances. Install flux (https://justgetflux.com) on your laptop to prevent your computer from disrupting your circadian rhythm. Aim to go to bed the same time each night and no hitting snooze on the alarm clock in the morning!

Revision Timetable

Still struggling to get organised? Then try filling in the example revision timetable below, remember to factor in enough time for short breaks, and stick to it! Remember to schedule in several breaks throughout the day and actually use them to do something you enjoy e.g. TV, reading, YouTube etc.

	8AM	10AM	12PM	2PM	4PM	6PM	8PM
MONDAY							
TUESDAY							
WEDNESDAY							
THURSDAY							
FRIDAY							
SATURDAY							
SUNDAY							
EXAMPLE DAY			School		Statistics	Pure Maths	Critical Thinking

Top tip! Ensure that you take a watch that can show you the time in seconds into the exam. This will allow you have a much more accurate idea of the time you're spending on a question. In general, if you've spent >150 seconds on a section 1 question – move on regardless of how close you think you are to solving it.

Getting the most out of Mock Papers

Mock exams can prove invaluable if tackled correctly. Not only do they encourage you to start revision earlier, they also allow you to **practice and perfect your revision technique**. They are often the best way of improving your knowledge base or reinforcing what you have learnt. Probably the best reason for attempting mock papers is to familiarise yourself with the exam conditions of the ECAA as they are particularly tough.

Start Revision Earlier

Thirty five percent of students agree that they procrastinate to a degree that is detrimental to their exam performance. This is partly explained by the fact that they often seem a long way in the future. In the scientific literature this is well recognised, Dr. Piers Steel, an expert on the field of motivation states that *'the further away an event is, the less impact it has on your decisions'*.

Mock exams are therefore a way of giving you a target to work towards and motivate you in the run up to the real thing – every time you do one treat it as the real deal! If you do well then it's a reassuring sign; if you do poorly then it will motivate you to work harder (and earlier!).

Practice and perfect revision techniques

In case you haven't realised already, revision is a skill all to itself, and can take some time to learn. For example, the most common revision techniques including **highlighting and/or re-reading are quite ineffective** ways of committing things to memory. Unless you are thinking critically about something you are much less likely to remember it or indeed understand it.

Mock exams, therefore allow you to test your revision strategies as you go along. Try spacing out your revision sessions so you have time to forget what you have learnt in-between. This may sound counterintuitive but the second time you remember it for longer. Try teaching another student what you have learnt, this forces you to structure the information in a logical way that may aid memory. Always try to question what you have learnt and appraise its validity. Not only does this aid memory but it is also a useful skill for Oxbridge interviews and beyond.

Improve your knowledge

The act of applying what you have learnt reinforces that piece of knowledge. A question may ask you to think about a relatively basic concept in a novel way (not cited in textbooks), and so deepen your understanding. Exams rarely test word for word what is in the syllabus, so when running through mock papers try to understand how the basic facts are applied and tested in the exam. As you go through the mocks or past papers take note of your performance and see if you consistently under-perform in specific areas, thus highlighting areas for future study.

Get familiar with exam conditions

Pressure can cause all sorts of trouble for even the most brilliant students. The ECAA is a particularly time pressured exam with high stakes – your future (without exaggerating) does depend on your result to a great extent. The real key to the ECAA is overcoming this pressure and remaining calm to allow you to think efficiently.

Mock exams are therefore an excellent opportunity to devise and perfect your own exam techniques to beat the pressure and meet the demands of the exam. **Don't treat mock exams like practice questions – it's imperative you do them under time conditions.**

> ***Remember!*** It's better that you make all the mistakes you possibly can now in mock papers and then learn from them so as not to repeat them in the real exam.

Before using this Book

Do the ground work

- Read in detail: the background, methods, and aims of the ECAA as well logistical considerations such as how to take the ECAA in practice. A good place to start is a ECAA textbook like *The Ultimate ECAA Guide* (flick to the back to get a free copy!) which covers all the groundwork.
    - It is generally a good idea to start re-capping all your GCSE and AS maths.
- Remember that calculators are not permitted in the exam, so get comfortable doing more complex long addition, multiplication, division, and subtraction.
- Get comfortable rapidly converting between percentages, decimals, and fractions.
- Practice developing logical arguments and structuring essays with an obvious introduction, main body, and ending.
- These are all things which are easiest to do alongside your revision for exams before the summer break. Not only gaining a head start on your ECAA revision but also complimenting your year 12 studies well.
- Discuss topical economics problems with others - propose theories and be ready to defend your argument. This will rapidly build your scientific understanding for section 2 but also prepare you well for an oxbridge interview.
- Read through the ECAA syllabus before you start tackling whole papers. This is absolutely essential. It contains several stated formulae, constants, and facts that you are expected to apply - or may just be an answer in their own right. Familiarising yourself with the syllabus is also a quick way of teaching yourself the additional information other exam boards may learn which you do not. Sifting through the whole ECAA syllabus is a time-consuming process so we have done it for you. **Be sure to flick through the syllabus checklist** later on, which also doubles up as a great revision aid for the night before!

Ease in gently

With the ground work laid, there's still no point in adopting exam conditions straight away. Instead invest in a beginner's guide to the ECAA, which will not only describe in detail the background and theory of the exam, but take you through section by section what is expected. *The Ultimate ECAA Guide* is the most popular ECAA textbook – you can get a free copy by flicking to the back of this book.

When you are ready to move on to past papers, take your time and puzzle your way through all the questions. Really try to understand solutions. A past paper question won't be repeated in your real exam, so don't rote learn methods or facts. Instead, focus on applying prior knowledge to formulate your own approach.

If you're really struggling and have to take a sneak peek at the answers, then practice thinking of alternative solutions, or arguments for essays. It is unlikely that your answer will be more elegant or succinct than the model answer, but it is still a good task for encouraging creativity with your thinking. Get used to thinking outside the box!

Accelerate and Intensify

Start adopting exam conditions after you've done two past papers. Don't forget that **it's the time pressure that makes the ECAA hard** – if you had as long as you wanted to sit the exam you would probably get 100%. If you're struggling to find comprehensive answers to past papers then ECAA *Past Papers Worked Solutions* contains detailed explained answers to every ECAA past paper question and essay (flick to the back to get a free copy).

Doing every past paper at least twice is a good target for your revision. In any case, choose a paper and proceed with strict exam conditions. Take a short break and then mark your answers before reviewing your progress. For revision purposes, as you go along, keep track of those questions that you guess – these are equally as important to review as those you get wrong.

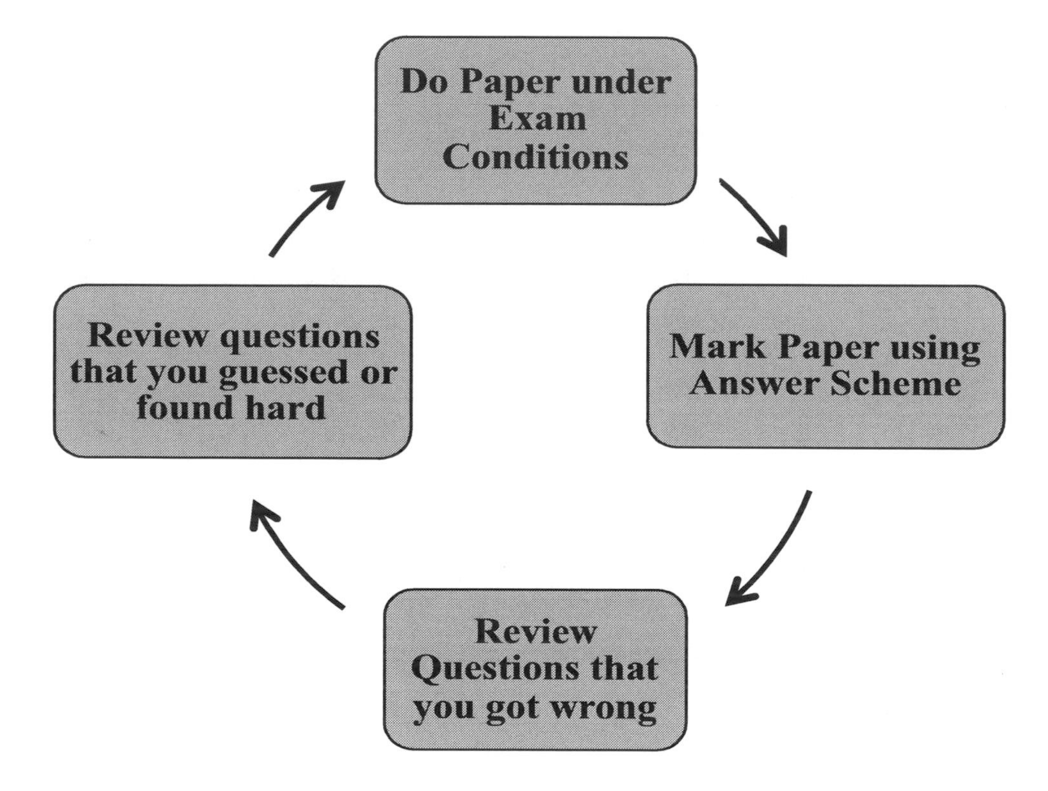

Once you've exhausted all the past papers, move on to tackling the unique mock papers in this book. In general, you should aim to complete one to two mock papers every night in the ten days preceding your exam.

Section 1A: An Overview

What will you be tested on?	No. of Questions	Duration
Problem-solving skills, numerical and spatial reasoning, critical thinking skills, understanding arguments and reasoning	20 MCQs	80 Minutes (incl. Section 1B)

This is the first section of the ECAA, comprising a total of 20 MCQ questions. You have 90 minutes in total to complete section 1, including the maths part (section 1B). It's best to devote 40 minutes to each subsection which gives you approximately 2 minutes for each question.

Not all the questions are of equal difficulty and so as you work through the past material it is certainly worth learning to recognise quickly which questions you should spend less time on in order to give yourself more time for the trickier questions.

Deducing arguments

Several MCQ questions will be aimed at testing your understand of the writer's argument. It is common to see questions asking you 'what is the writer's view?' or 'what is the writer trying to argue?'. This is arguably an important skill you will have to develop, and the TSA is designed to test this ability. You have limited time to read the passage and understand the writer's argument, and the only way to improve your reading comprehension skill is to read several well-written news articles on a daily basis and think about them in a critical manner.

Assumptions

It is important to be able to identify the assumptions that a writer makes in the passage, as several questions might question your understand of what is assumed in the passage. For example, if a writer mentions that 'if all else remains the same, we can expect our economic growth to improve next year', you can identify an assumption being made here – the writer is clearly assuming that all external factors remain the same.

Fact vs. Opinion

It is important to **be able to decipher whether the writer is stating a fact or an opinion** – the distinction is usually rather subtle and you will have to decide whether the writer is giving his or her own personal opinion, or presenting something as a fact. Section 1 may contain questions that will test your ability to identify what is presented as a fact and what is presented as an opinion.

Fact	Opinion
'There are 7 billion people in this world...'	'I believe there are more than 7 billion people in this world...'
'She is an Australian...'	'She sounded like an Australian...'
'Trump is the current President...'	'Trump is a horrible President...'
'Vegetables contain a lot of fibre...'	'Vegetables are good for you...'

Numerical and spatial reasoning

There are several questions that will test how well you can cope with numbers, and you should ideally be comfortable with simple mental calculations and being able to think logically.

Section 1B: An Overview

What will you be tested on?	No. of Questions	Duration
The ability to apply mathematical knowledge up to A Level	15 MCQs	80 Minutes (incl. Section 1A)

Section 1B of the exam involves short MCQ questions relating to Mathematics that are designed to see if you can quickly apply the principles that you have learnt in school in a time pressured exam. Assuming you split your time evenly between sections 1A + 1B (40 minutes each), you will have on average 160 seconds per question so it vital to work very quickly- some questions later tend to be harder so you should be doing the initial questions in under 90 seconds. I cannot emphasise enough that the limiting factor in this test is time not your ability. Practice is therefore crucial to learn the technique, skills and tricks to answer section 1 questions quickly. A quick summary of the syllabus is included below:

- **Number**- you should be confident performing a wide range of numerical calculations without the use of a calculator. As this is a MCQ exam, producing order of magnitude estimates will be very useful
- **Algebra & Functions**- you should be competent at basic algebra taught up to AS maths. You will already be at this standard but the key is to practice lots of questions so you use your algebra at the required speed The inequalities can be challenging to do under to time pressure so we recommend quickly drawing out the xy plane and identifying the region of interest. The factor/remainder theorem from A2 maths also appear in the syllabus so you may be tested on this.
- **Measure**- this is linked to "number" but we recommend that you become fully confident when dealing with scale factors. A question can often be simplified by working with this approach.
- **Statistics**- a very basic knowledge of GCSE statistics is all that is necessary. It is however important to know how to combine different statistics together and not get bogged down in long calculations
- **Probability**- a basic GCSE level of knowledge of probability but you will need to work through these questions quickly. We recommend that you practice drawing out 'tree diagrams' quickly to solve these problems
- **Coordinate Geometry** in the (x,y) plane- This is also content covered in AS Level maths and the challenge will be completing questions under time pressure. Practise converting equations to a standard form and then sketching them on the xy plane- this will often help you spot the solution. As this is a MCQ exam, you will not need to provide geometric proofs.
- **Trigonometry** - Basic trigonometry covering material tested in AS level maths. You are expected to know two basic trig formulae as well as the values of sine, cosine and tangent for the angles 0°, 30°, 45°, 60°, 90°. As you will not have a calculator, it is crucial to memorise these values. This will also be very useful for interviews.
- **Exponential and Logarithms**- be confident at using the log formulae that you learnt in AS level maths. Using the formula will often simplify a question and with practice you will be able to determine whether to solve an equation in exponential form or logarithmic form
- **Differentiation**- when sitting your ENGAA exam, you will likely have covered advanced topics in differentiation including the product, chain and quotient rule. However, the exam itself only tests very basic differentiation taught in AS maths so try not to overcomplicate these questions.
- **Integration**- this once again contains the basic integration taught as part of AS level maths. It will be important to practice definite integration without a calculator as it is very easy to make a simple mistake.
- **Graphs of Functions**- this is a very important topic as it provides a lot of tricks to solve maths questions. We recommend you know the C3 transformations of graphs inside out and draw out sketches of common functions

Section 2: An Overview

What will you be tested on?	No. of Questions	Duration
Your ability to write an essay under timed conditions, your writing technique and your argumentative abilities	No Choice – Only one question	40 Minutes

Section 2 is usually what students are more comfortable with – after all, many GCSE and A Level subjects require you to write essays within timed conditions. It does not require you to have any particular legal knowledge – the questions can be very broad and cover a wide range of topics.

Here are some of the topics that might appear in Section two:

- Science
- Politics
- Religion
- Technology
- Ethics
- Morality
- Philosophy
- Education
- History
- Geopolitics

As you can see, this list is very broad and definitely non-exhaustive, and you do not get many choices to choose from (you have to write one essay out of three choices). Many students make the mistake of focusing too narrowly on one or two topics that they are comfortable with – this is a dangerous gamble and if you end up a topic you are unfamiliar with, this is likely to negatively impact your score.

You should ideally focus on at least four topics to prepare from the ECAA, and you can pick and choose which topics from the list above are the ones you would be more interested in. Here are some suggestions:

Economics Science

An essay that is related to science might relate to recent technological advancements and their implications, such as the rise of Bitcoin and the use of blockchain technology and artificial intelligence. This is interrelated to ethical and moral issues, hence you cannot merely just regurgitate what you know about artificial intelligence or blockchain technology. The examiners do not expect you to be an expert in an area of science – what they want to see is how you identify certain moral or ethical issues that might arise due to scientific advancements, and how do we resolve such conundrums as human beings.

Politics

Politics is undeniably always a hot topic and consequently a popular choice amongst students. The danger with writing a politics question is that some students get carried away and make their essay too one-sided or emotive – for example a student may chance upon an essay question related to Brexit and go on a long rant about why the referendum was a bad idea. You should always remember to answer the question and make sure your essay addresses the exact question asked – do not get carried away and end up writing something irrelevant just because you have strong feelings about a certain topic.

Religion

Religion is always a thorny issue and essays on religion provide strong students with a good opportunity to stand out and display their maturity in thought. Questions can range from asking about your opinion with regards to banning the wearing of a headdress to whether children should be exposed to religious practices at a young age. Questions related to religion will require a student to be sensitive and measured in their answers and it is easy to trip up on such questions if a student is not careful.

Education

Education is perhaps always a relatable topic to students, and students can draw from their own experience with the education system in order to form their opinion and write good essays on such topics. Questions can range from whether university places should be reduced, to whether we should be focusing on learning the sciences as opposed to the arts.

Section 2: Revision Guide

SCIENCE

Resource	What to read/do
1. **Newspaper Articles**	• The Guardian, The Times, The Economist, The Financial Times, The Telegraph, The New York Times, The Independent
2. **A Levels/IB**	• Look at the content of your science A Levels/IB if you are doing science subjects and critically analyse what are the potential moral/ethical implications • Use your A Levels/IB resources in order to seek out further readings – e.g. links to a scientific journal or blog commentary • Remember that for your LNAT essay you should not focus on the technical issues too much – think more about the ethical and moral issues
3. **Online videos**	• There are plenty of free resources online that provide interesting commentary on science and the moral and ethical conundrums that scientists face on a daily basis • E.g. Documentaries and specialist science channels on YouTube • National Geographic, Animal Planet etc. might also be good if you have access to them
4. **Debates**	• Having a discussion with your friends about topics related to science might also help you formulate some ideas • Attending debate sessions where the topic is related to science might also provide you with excellent arguments and counter-arguments • Some universities might also host information sessions for sixth form students – some might be relevant to ethical and moral issues in science
5. **Museums**	• Certain museums such as the Natural Science Museum might provide some interesting information that you might not have known about
6. **Non-fiction books**	• There are plenty of non-fiction books (non-technical ones) that might discuss moral and ethical issues about science in an easily digestible way

POLITICS

Resource	What to read/do
1. **Newspaper Articles**	• The Guardian, The Times, The Economist, The Financial Times, The Telegraph, The New York Times, The Independent
2. **Television**	• Parliamentary sessions • Prime Minister Questions • Political news
3. **Online videos**	• Documentaries • YouTube Channels
4. **Lectures**	• University introductory lectures • Sixth form information sessions
5. **Debates**	• Debates held in school • Joining a politics club
6. **Podcasts**	• Political podcasts • Listen to both sides to get a more rounded view (e.g. listening to both left and right wing podcasts)

RELIGION

Syllabus Point	What to read/do
1. **Newspaper Articles**	• The Guardian, The Times, The Economist, The Financial Times, The Telegraph, The New York Times, The Independent
2. **Non-fiction books**	• Read up about books that explain the origins and beliefs of different types of religion • E.g. Books that talk about the origins of Christianity, Islam or Buddhism, theology books etc.
3. **Talking to religious leaders**	• Talking to religious leaders may be a good way of understanding different religions more and being able to write an essay on religion with more maturity and nuance • Talking to people from different religious backgrounds may also be a good way of forming a more well-rounded opinion
4. **Online videos**	• Documentaries on religion • YouTube channels providing informative and educational videos on different religions – e.g. history, background
5. **Lectures**	• Information sessions • Relevant introductory lectures
6. **Opinion articles**	• Informative blogs and journals • Read both arguments and counter-arguments and come up with your own viewpoint

EDUCATION

Syllabus Point	What to read/do
1. **Newspaper Articles**	• The Guardian, The Times, The Economist, The Financial Times, The Telegraph, The New York Times, The Independent
2. **A Levels/IB**	• Draw inspiration from what you are studying in your A Levels or IB – do you feel like what you are studying is useful and relevant? E.g. Studying arts versus science • Compare the education you are receiving with your friends in different schools or different subjects
3. **Educational exchange**	• If you have an opportunity to go on an educational exchange, this might be a good opportunity to compare and contrast different educational systems • E.g. the approach to education in Germany versus the UK
4. **University applications**	• Have a read of how different universities promote themselves – do they claim to provide students with academic enlightenment, or better job prospects, or a good social life? • Why do different universities focus on different things?
5. **Online videos**	• Documentaries • YouTube Channels
6. **Talk to your teachers**	• Your teachers have been in the education industry for years and maybe decades – talk to them and ask them for their opinion • Talk to different teachers and compare their opinions regarding how we should approach education

Maths Revision Checklist

The material for the overviews of sections one and two have mainly been taken from the 2017 syllabus - this may change in the future. We recommend you consult the most up to date syllabus to see if there are any differences.

Syllabus Point	What to Know
1. Number	Understand and use BIDMAS Define; factor, multiple, common factor, highest common factor, least common multiple, prime number, prime factor decomposition, square, positive and negative square root, cube and cube root Use index laws to simplify multiplication and division of powers Interpret, order and calculate with numbers written in standard index form Convert between fractions, decimals and percentages Understand and use direct and indirect proportion; Apply the unitary method Use surds and π in exact calculations, simplify expressions that contain surds. Calculate upper and lower bounds to contextual problems Rounding to a given number of decimal places or significant figures
2. Algebra	Simplify rational expressions by cancelling or factorising and cancelling Set up quadratic equations and solve them by factorising Set up and use equations to solve problems involving direct and indirect proportion Use linear expressions to describe the nth term of a sequence Use Cartesian coordinates in all four quadrants Equation of a straight line, $y=mx+c$, parallel lines have the same gradient Graphically solve simultaneous equations Recognise and interpret graphs of simple cubic functions, the reciprocal function, trigonometric functions and the exponential function $y=kx$ for integer values of x and simple positive values of k Draw transformations of $y=f(x)$ **[**$(y=af(x)$, $y=f(ax)$, $y=f(x)+a$, $y=f(x-a)$ **only]**
3. Geometry	Recall and use properties of angle at a point, on a straight line, perpendicular lines and opposite angles at a vertex, and the sums of the interior and exterior angles of polygons Understand congruence and similarity; Use Pythagoras' theorem in 2-D and 3-D Use the trigonometric ratios, between 0° and 180°, to solve problems in 2-D and 3-D Understand and construct geometrical proofs, including using circle theorems: a. **the angle subtended at the circumference in a semicircle is a right angle** b. **the tangent at any point on a circle is perpendicular to the radius at that point** Describe and transform 2-D shapes using single or combined rotations, reflections, translations, or enlargements, including the use of vector notation
4. Measures	Calculate perimeters and areas of shapes made from triangles, rectangles, and other shapes, find circumferences and areas of circles, including arcs and sectors Calculate the volumes and surface areas of prisms, pyramids, spheres, cylinders, cones and solids made from cubes and cuboids (formulae given for the sphere and cone) Use vectors, including the sum of two vectors, algebraically and graphically Discuss the inaccuracies of measurements; Understand and use three-figure bearings
5. Statistics	Identify possible sources of bias in experimental methodology Discrete vs. continuous data; Design and use two-way tables Interpret cumulative frequency tables and graphs, box plots and histograms Define mean, median, mode, modal class, range, and inter-quartile range Interpret scatter diagrams and recognise correlation, drawing and using lines of best fit Compare sets of data by using statistical measures
6. Probability	List all the outcomes for single and combined events Identify mutually exclusive outcomes; the sum of the probabilities of all these outcomes is 1 Construct and use Venn diagrams Know when to add or multiply two probabilities, and understand conditional probability Understand the use of tree diagrams to represent outcomes of combined events

How to use Practice Papers

If you have done everything this book has described so far then you should be well equipped to meet the demands of the ECAA, and therefore **the mock papers in the rest of this book should ONLY be completed under exam conditions**.

This means:

- Absolute silence – no TV or music
- Absolute focus – no distractions such as eating your dinner
- Strict time constraints – no pausing half way through
- No checking the answers as you go
- Give yourself a maximum of three minutes between sections – keep the pressure up
- Complete the entire paper before marking
- Mark harshly

In practice this means setting aside two hours in an evening to find a quiet spot without interruptions and tackle the paper. Completing one mock paper every evening in the week running up to the exam would be an ideal target.

- Tackle the paper as you would in the exam.
- Return to mark your answers, but mark harshly if there's any ambiguity.
- Highlight any areas of concern.
- If warranted read up on the areas you felt you underperformed to reinforce your knowledge.
- If you inadvertently learnt anything new by muddling through a question, go and tell somebody about it to reinforce what you've discovered.

Finally relax… the ECAA is an exhausting exam, concentrating so hard continually for two hours will take its toll. So, being able to relax and switch off is essential to keep yourself sharp for exam day! Make sure you reward yourself after you finish marking your exam.

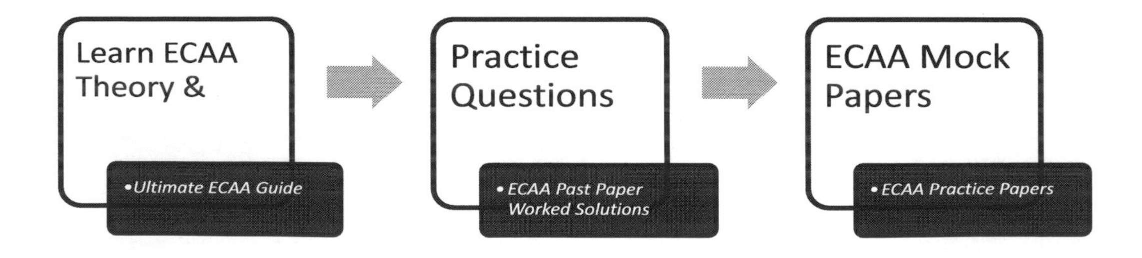

Mock Paper A

Section 1A

Question 1

Competitors need to be able to run 200 metres in under 25 seconds to qualify for a tournament. James, Steven and Joe are attempting to qualify. Steven and Joe run faster than James. James' best time over 200 metres is 26.2 seconds. Which response is definitely true?

A. Only Joe qualifies
B. James does not qualify.
C. Joe and Steven both qualify
D. Joe qualifies
E. No one qualifies

Question 2

You spend £5.60 in total on a sandwich, a packet of crisps and a watermelon. The watermelon cost twice as much as the sandwich, and the sandwich cost twice the price of the crisps. How much did the watermelon cost?

A. £1.20
B. £2.60
C. £2.80
D. £3.20
E. £3.60

Question 3

Jane, Chloe and Sam are all going by train to a football match. Chloe gets the 2:15pm train. Sam's journey takes twice as long Jane's. Sam catches the 3:00pm train. Jane leaves 20 minutes after Chloe and arrives at 3:25pm. When will Sam arrive?

A. 3:50pm
B. 4:10pm
C. 4:15pm
D. 4:30pm
E. 4:40pm

Question 4

Michael has eleven sweets. He gives three sweets to Hannah. Hannah now has twice the number of sweets Michael has remaining. How many sweets did Hannah have before the transaction?

A. 11
B. 12
C. 13
D. 14
E. 15

Question 5

Alex gets a pay rise of 5% plus an extra £6 per week. The flat rate of income tax is decreased from 14% to 12% at the same time. Alex's current weekly take-home pay is £250 per week.
What will his new weekly take-home pay be, to the nearest whole pound?

A. £260
B. £267
C. £273
D. £279
E. £285

MOCK PAPER A SECTION 1A

Question 6

You have four boxes, each containing two coloured cubes. Box A contains two white cubes, Box B contains two black cubes, and Boxes C and D both contain one white cube and one black cube. You pick a box at random and take out one cube. It is a white cube. You then draw another cube from the same box.

What is the probability that this cube is not white?

A. ½ B. ⅓ C. ⅔ D. ¼ E. ¾

Question 7

Anderson & Co. hire out heavy plant machinery at a cost of £500 per day. There is a surcharge for heavy usage, at a rate of £10 per minute of usage over 80 minutes. Concordia & Co. charge £600 per day for similar machinery, plus £5 for every minute of usage.

For what duration of usage are the costs the same for both companies?

A. 100 minutes B. 130 minutes C. 140 minutes D. 170 minutes E. 180 minutes

Question 8

Simon is discussing with Seth whether or not a candidate is suitable for a job. When pressed for a weakness at interview, the candidate told Simon that he is a slow eater. Simon argues that this will reduce the candidate's productivity, since he will be inclined to take longer lunch breaks.

Which statement **best** substantiates Simon's argument?

A. Slow eaters will take longer to eat lunch
B. Longer lunch breaks are a distraction
C. Eating more slowly will reduce the time available to work
D. Eating slowly is a weakness
E. People who like food are more likely to eat slowly

Question 9

Three pieces of music are on repeat in different rooms of a house. One piece of music is three minutes long, one is four minutes long and the final one is 100 seconds long. All pieces of music start playing at exactly the same time. How long is it until they are next all starting together?

A. 12 minutes B. 15 minutes C. 20 minutes D. 60 minutes E. 300 minutes

Question 10

A car leaves Salisbury at 8:22am and travels 180 miles to Lincoln, arriving at 12:07pm. Near Warwick, the driver stopped for a 14 minute break. What was its average speed, whilst travelling, in kilometres per hour? It should be assumed that the conversion from miles to kilometres is 1:1.6.

A. 51kph B. 67kph  C. 77kph D. 82kph E. 386kph

Questions **11** and **12** refer to the following data:

Five respondents were asked to estimate the value of three bottles of wine, in pounds sterling.

Respondent	Wine 1	Wine 2	Wine 3
1	13	16	25
2	17	16	23
3	11	17	21
4	13	15	14
5	15	19	29
Actual retail value	**8**	**25**	**23**

Question 11

What is the mean error margin in the guessing of the value of wine 1?

A. £4.80 B. £5.60 C. £5.80 D. £6.20 E. £6.40

Question 12

Which respondent guessed most accurately on average?

A. Respondent 1 B. Respondent 2 C. Respondent 3 D. Respondent 4 E. Respondent 5

Questions **13** and **14** refer to the following data:

The population of Country A is 40% greater than the population of Country B.

The population of Country C is 30% less than the population of Country D (which is has a population 20% greater than Country B).

Question 13

Given that the population of Country A is 45 million, what is the population of country D?

A. 32.1 million B. 35.8 million C. 36.6 million D. 38.6 million E. 39.0 million

Question 14

The population of Country A is still 45 million. If Country B introduced a new health initiative costing $45 per capita, what would be the total cost?

A. $1.35 bn B. $1.45 bn C. $1.50 bn D. $1.55 bn E. $1.65 bn

Question 15

A car averages a speed of 30mph over a certain distance and then returns over the same distance at an average speed of 20mph. What is the average speed for the journey as a whole?

A. 22.5 mph B. 24 mph C. 25 mph D. 26 mph

E. The distance travelled is required to calculate average speed

Question 16

"All sheep are ruminants and all marsupials are mammals. No sheep are marsupials." Which of the following must be true?

A. Some ruminants are marsupials. B. All mammals are marsupials C. All sheep are mammals

D. Some sheep are marsupials. E. None of the above

Question 17

The price of toothpaste rises by 80%. This is later reduced by 50% due to competition. Zoe buys two tubes of toothpaste and gets the third free because of a loyalty card. How much did she have to pay per tube of toothpaste? Express your answer as a percentage of the original price.

A. 16.67% B. 33% C. 60% D. 66.7% E. 100%

Question 18

"You can remain fit throughout life if you exercise regularly. Simon does not exercise regularly, so he can never become fit." Which flawed argument has the same structure as this?

A. "You can speak a foreign language if you learn when young. Simon does not speak a foreign language, so he did not learn when young."

B. "You are never tired if you sleep for 8 hours a night. Simon is tired, therefore he doesn't sleep for 8 hours a night"

C. "You can be a good musician if you practice regularly. Simon does not practice regularly, so he can never be a good musician."

D. "You can be good at sport if you have a natural ability. Simon is good at hockey, therefore he has a natural ability."

E. "Eating five portions of fruit and vegetables daily reduces the risk of heart disease. Simon eats more than this, so he will not develop heart disease."

Question 19

"Reports of cybercrime are increasing year on year. Last year, police dealt with 250% more cybercrime then the year before. Common complaints relate to inappropriate or defamatory use of social media. To deal with this, many police forces are creating dedicated teams to deal with online offences. A pilot study showed that a dedicated cybercrime team solved cases of cybercrime 40% faster than regular detectives. Therefore the measure will act to suppress the rise in cybercrime."

Which statement best validates the above argument?

A. Solving crimes faster is necessary to keep pace with the increase in crime
B. Solving crimes faster leads to more convictions
C. Solving crimes faster increases police resources to tackle crime
D. Solving crimes faster saves money
E. Solving crimes faster reassures the public of action

Question 20

"Recently in Kansas, a number of farm animals have been found killed in the fields. The nature of the injuries is mysterious, but consistent with tales of alien activity. Local people talk of a number of UFO sightings, and claim extra terrestrial responsibility. Official investigations into these claims have dismissed them, offering rational explanations for the reported phenomena. However, these official investigations have failed to deal with the point that, even if the UFO sightings can be explained in rational terms, the injuries on the carcasses of the farm animals cannot be. Extra terrestrial beings must therefore be responsible for these attacks."

Which of the following best expresses the main conclusion of this argument?

A. Sightings of UFOs cannot be explained by rational means
B. Recent attacks must have been carried out by extraterrestrial beings
C. The injuries on the carcasses are not due to normal predators
D. UFO sightings are common in Kansas
E. Official investigations were a cover-up

Question 21

"To make a cake you must prepare the ingredients and then bake it in the oven. You purchase the required ingredients from the shop, however the oven is broken. Therefore you cannot make a cake."

Which of the following arguments has the same structure?

A. To get a good job, you must have a strong CV then impress the recruiter at interview. Your CV was not as good as other applicants, therefore you didn't get the job.

B. To get to Paris, you must either fly or take the Eurostar. There are flight delays due to dense fog, therefore you must take the Eurostar.

C. To borrow a library book, you must go to the library and show your library card. At the library, you realise you have forgotten your library card. Therefore you cannot borrow a book.

D. To clean a bedroom window, you need a ladder and a hosepipe. Since you don't have the right equipment, you cannot clean the window.

E. Bears eat both fruit and fish. The river is frozen, so the bear cannot eat fish.

Question 22

"Growing vegetables requires patience, skill and experience. Patience and skill without experience is common – but often such people give up prematurely as skill alone is insufficient to grow vegetables, and patience can quickly be exhausted."

Which of the following summarises the main argument?

A. Most people lack the skill needed to grow vegetables

B. Growing vegetables requires experience

C. The most important thing is to get experience

D. Most people grow vegetables for a short time but give up due to a lack of skill

E. Successful vegetable growers need to have several positive traits

Section 1B

Question 23

If the lines $y_1 = (n + 1)x + 10$and $y_2 = (n + 3)x + 2$ are perpendicular then n must equal which of the following?

A. 2 B. -2 C. 3 D. -3 E. 0 F. 1

Question 24

The curve $y = x^2 + 3$ is reflected about the line $y = x$ and subsequently translated by the vector $\begin{pmatrix} 4 \\ 2 \end{pmatrix}$. Which of the following is the x-intercept of the resulting curve?

A. -2 B. 11 C. 7 D. -11 E. 8 F. -8

Question 25

Given that $a^{3x}b^xc^{4x} = 2$, where a > 0, b > 0, and c > 0, then x =

A. $\frac{2}{3a + b + 4c}$

B. $\frac{\log_{10}2}{\log_{10}(a^3bc^4)}$

C. $\frac{log_{10}2}{log_{10}(a^3bc^4)}$

D. $log_{10}\frac{2}{(ab^2c^3)}$

E. $\frac{log_2 10}{log_2(a^3bc^4)}$

F. $log_{10}\frac{2}{(a^3bc^4)}$

Question 26

The sum of the roots of the equation $2^{2x} - 8 \times 2^x + 15 = 0$ is

A. 4

B. 16

C. $log_{10}\left(\frac{15}{2}\right)$

D. $\frac{log_{10}15}{log_{10}2}$

E. 8

F. $log_2\left(\frac{2}{3}\right)$

Question 27

For what values of the non-zero real number a does the equation $a x^2 + (a - 2)x = 2$ have real and distinct roots?

A. $a \neq -2$

B. $a > 2$

C. $a > -2$

D. No values of a.

E. $a \neq 0$

F. $a > 5$

Question 28

A bag only contains 2n blue balls and n red balls. All the balls are identical apart from colour. One ball is randomly selected and not replaced. A second ball is then randomly selected. What is the probability that at least one of the selected balls is red?

A.	$\frac{4n}{3(3n-1)}$	B.	$\frac{5n-1}{3(3n-1)}$	C.	$\frac{5n-5}{3(3n-1)}$
D.	$\frac{4n-2}{3(3n-1)}$	E.	$\frac{n-5}{9(n-1)}$	F.	$\frac{4n-1}{3(3n-1)}$

Question 29

Which of the following equations is a correct simplification of the equation $\frac{x^2-16}{x^2-4x}$?

A.	$1-\frac{4}{x}$	B.	$\frac{x+4}{x}$	C.	$\frac{x-4}{x}$
D.	$\frac{4}{x}$	E.	$\frac{x(x-4)}{x}$	F.	$\frac{x+4}{4x}$

Question 30

What is the equation of the quadratic function that passes through the x-coordinates of the stationary points of $y = x^2e^x$?

A.	Function does not exist.	B.	x^2+2x	C.	x^2
D.	x^2-2x	E.	x^2+4x	F.	$2x^2-1$

Question 31

Given a curve with the equation $y = 8 - 4x - 2x^2$ and a line $y = k(x+4)$, find the values of k for which the line and the curve are tangent to each other.

A.	$-4 < k \leq 4$	B.	$k = 4, k = 20$	C.	$4 < k < 20$	D.	$k = -4, k = 4$

Question 32

Given the two equations $y_1 = (1-x)^6$ and $y_2 = (1+2x)^6$, find the ratio of the coefficients of the 2nd term in the expansion of y_1 and the 3rd term in the expansion of y_2 (The y_1 coefficient should be the numerator, and the y_2 coefficient should be the denominator).

A.	$\frac{-1}{10}$	B.	$\frac{1}{9}$	C.	$\frac{1}{15}$	D.	$-\frac{1}{7}$.

Question 33

What is the sum of the integers from 1 to 300?

A. 9,000 B. 44,850 C. 45,150 D. 45,450 E. 54,450 F. 90,000

Question 34

If $sin2\theta = \frac{2}{5}$, then what is $\frac{1}{sin\theta cos\theta}$?

A. $\frac{1}{5}$ B. $\frac{5}{4}$ C. $\frac{5}{2}$ D. 5 E. $\frac{3}{2}$ F. 1

Question 35

If $\lceil n \rceil$ represents the greatest integer less than or equal to n, then which of the following is the solution to $-11 + 4\lceil n \rceil = 5$?

A. n = 4 B. 4<n<5 C. -2≤n≤-1 D. 4≤n<5 E. 4<n≤5 F. n<5

Question 36

The operation ∅ is defined for all real numbers a and b$(b \neq 0)$ as $a \varnothing b = \frac{a/2}{b}$.

If $10 \varnothing n = n \varnothing \left(\frac{1}{10}\right)$, which of the following is a solution for n?

A. 1 B. $5\sqrt{2}$ C. -10
D. 10 E. $-5\sqrt{2}$ F. -1

Question 37

If -1 is a zero of the function f(x) = $2x^3 + 3x^2 - 20x - 21$, then what are the other zeroes?

A. 1 and 3 B. -3 and 3 C. $\frac{-7}{2}$ and 1 and 3
D. $\frac{-7}{2}$ and 3 E. $-1 \wedge 3$ F. 1 and 7

Section 2

Read the extract taken from Neal Reaich's *What price honesty?* (2014, Bized) and then answer the question below in the space provided in this booklet.

Your answer will be assessed taking into account your ability to construct a reasoned, insightful and logically consistent argument with clarity and precision.

QUESTION
To what extent can prices be considered a true reflection of value, and to what extent should governments regulate price-setting behaviours such as reference pricing? Discuss with reference to the passage above.

What price honesty?

Pricing seems to be mentioned quite a lot in recent news headlines. The latest is the investigation by the Office of Fair Trading (OFT) into just how genuine the advertised price cuts in some large furniture stores really are. The OFT use the term reference pricing. They've found cases in shops under investigation where not a single product had, in reality, been sold at the, supposedly original, higher price. The argument goes that since 95% of sales were at the lower or 'now' price then the stated original prices cannot be really genuine.
Are we really so stupid as to be misled by reference pricing? When my wife buys some clothes in a sale and I ask how much it cost, she always responds by saying how much money she has saved. Well, unless she had the definite intention of buying it in the first place, she hasn't saved anything: in fact she has spent it. The question is whether or not the reference pricing made a difference to her decision to buy the product. It must do, otherwise why is it so commonly used?

Heuristics suggests that people are not as rational as the standard economic model implies. Instead they use rules of the thumb, educated guesses or short cuts in decision-making. Anchoring refers to people making decisions based upon something they know to start with. For example the anchor price of a jumper was £40 and there is 20 per cent discount offer making the new price £32 a saving of £8. If I decide to buy I might be thinking of the £8 saved because I didn't have to pay the full price, when I should be thinking logically about the actual price I am paying. In this case there is too much focus on the anchor price, which is at a level that I wouldn't have purchased the item anyway.

In some cases reference pricing is part of a strategy of price discrimination over time. Clothes shops attempt to capture consumer surplus by charging a high initial price for a few weeks for the 'must have and will pay' customers. The shops then give a discount which increases over time. New potential customers must now weigh up whether to buy now and get a 25 per cent discount or wait longer for a 5per cent saving and risk losing the deal because they have run out of their size.

Another tactic is to use time-limited offers where notice is given that the offer ends soon: buy now or miss out. Double glazing sales were notorious for using this tactic, encouraging customers to 'sign up today to will get an extra 20 per cent off'.

Supermarkets are always giving volume offers, such as three for the price of two or get the second purchase for half the price. The supermarkets know we will pay more for the first item than we would for a second or third one. This tactic must lead to food waste as we are tempted to buy some products that we cannot possibly use before the sell-by date.

The OFT also look at baiting sales where only a very limited number of products are available at the most discounted price. This sounds a little like discounts given for advanced booking train tickets.

The OFT always makes mention of portioned 'drip' pricing where price increments drip through the buying process. It's those little add-ons that all add-up. Booking certain airline tickets comes to mind here.

Restaurants often make an offer of a free second main course meal after buying one main course at full price. Now add on the fact that we may have full priced deserts, order drinks at a high mark up and then add the tip. Get your calculator out and the deal doesn't seem so good.

All these examples of price framing seek to alter a consumer's perception of the value of the offer. But there is more: have you purchased a printer at a ridiculously low price only to be stung on purchasing printer ink cartridges? Some businesses operate at a loss or low profit margin on certain items, to entice customers to part with more money on higher-profit goods. They bundle items together.

Some advice then:

- Use a calculator with you when shopping
- When buying clothes divide the total by 50 to give you the average cost of wearing something once a week for a year: a shirt costing £25 will work out at 50 pence for every day worn;
- Avoid impulse buying on larger-ticket items;
- Only purchase multi buys with long sell buy dates;
- Ignore the original price: it's only the current price you need to know about;
- Avoid time limited offers.

END OF PAPER

Mock Paper B

Section 1A

Question 1
Joseph has a bag of building blocks of various shapes and colours. Some of the cubic ones are black. Some of the black ones are pyramid shaped. All blue ones are cylindrical. There is a green one of each shape. There are some pink shapes. Which of the following is definitely **NOT** true?

A. Joseph has pink cylindrical blocks
B. Joseph doesn't have pink cylindrical blocks
C. Joseph has blue cubic blocks
D. Joseph has a green pyramid
E. Joseph doesn't have a black sphere

Question 2
Sam notes that the time on a normal analogue clock is 1540hrs. What is the smaller angle between the hands on the clock?

A. 110° B. 120° C. 130° D. 140° E. 150°

Question 3
A fair 6-faced die has 2 sides painted red. The die is rolled 3 times. What is the probability that at least one red side has been rolled?

A. $8/27$ B. $19/27$ C. $21/27$ D. $24/27$ E. 1

Question 4
In a particular furniture warehouse, all chairs have four legs. No tables have five legs, nor do any have three. Beds have not less than four legs, but one bed has eight as they must have a multiple of four legs. Sofas have four or six legs. Wardrobes have an even number of legs, and sideboards have an odd number. No other furniture has legs. Brian picks a piece of furniture out, and it has six legs.
What can be deduced about this piece of furniture?

A. It is a table
B. It could be either a wardrobe or a sideboard
C. It must be either a table or a sofa
D. It must be either a table, a sofa or a wardrobe
E. It could be either a bed, a table or a sofa.

Question 5

Two friends live 42 miles away from each other. They walk at 3mph towards each other. One of them has a pet pigeon which starts to fly at 18mph as soon as the friends set off. The pigeon flies back and forth between the two friends until the friends meet. How many miles does the pigeon travel in total?

A. 63 B 84 C 114 D 126 E 252

Question 6

"Fruit juice contains fibre, vitamins and minerals and can be part of a healthy diet. However, it has been suggested that the high sugar content and acidity negates these benefits by leading to increased rates of dental cavities and hyperactivity in children. If left unchecked, a combination of poor dental hygiene and inappropriate diet can lead to disastrous consequences, including serious infections. On the other hand, many juices contain essential vitamins such as vitamin C which helps the immune system fight infections."

What is the main message from this passage?

A. Children should not drink fruit juice
B. Fruit juice is harmful to health
C. Fruit juice is good for health
D. On balance, we should drink more fruit juice
E. The overall benefits of fruit juice are unclear

Question 7

A complete stationery set includes a pen, a pencil, a geometry set and a pad of paper. Pens cost £1.50, pencils cost 50p, geometry sets cost £3 and paper pads cost £1. Sam, Dave and George each want complete sets, but Mr Browett persuades them to share. Sam and Dave agree to share a paper pad and a geometry set. George must have his own pen, but agrees that he and Sam can share a pencil.

What is the total amount spent?

A. £12.00 B. £13.50 C. £16.50 D. £17.50 E. £18.00

Question 8

"If the government financially supports the arts, a proportion of each person's taxes will be used to finance museums, galleries and theatres. But some taxpayers have no interest in the arts and never go to theatres or museums. Many of those who enjoy the arts are able to afford to pay for them. Since no one should be forced to subsidise services which they themselves do not use, taxpayers' money should not be used to support the arts."

Which counter-argument provides the strongest rebuke of this principle?

A. If public funding for the arts is withdrawn, only those who are genuinely interested would pay to visit museums

B. The rail network is publically subsidised, although some people do not use trains

C. If people only pay for services they use, then those who can afford private health insurance would not pay towards the NHS

D. Funding museums allows greater preservation of our heritage

E. If something requires subsidy, then people must not genuinely want it

Question 9

The figure to the right shows 5 squares made from 12 matches. Which 2 matches need to be moved to make 7 squares?

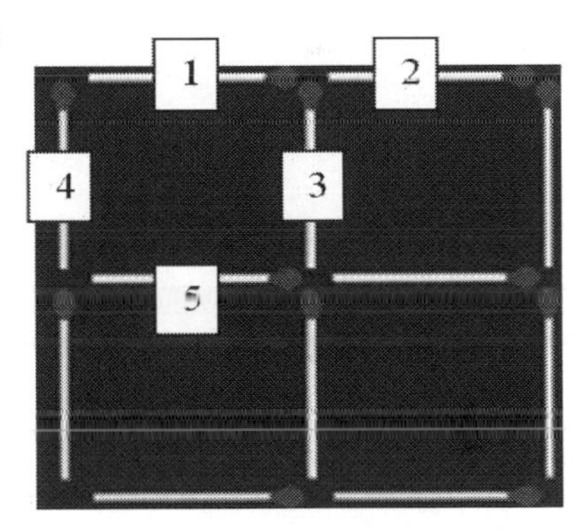

A. 1 and 2 B. 1 and 3 C. 1 and 4

D. 3 and 5 E. Not possible

Question 10

A cube has six sides of different colours. The red side is opposite to black. The blue side is adjacent to white. The brown side is adjacent to blue. The final side is yellow. Which colour is opposite brown?

A. Red B. Black C. Blue D. White E. Yellow

Question 11

The UK imports 36,000,000kg of cocoa beans each year. Each g costs the UK 0.3p, from which the supplier takes 20% commission. Of what is left, the local government takes 60% and the distribution company gets 30%. How much are the cocoa farmers left with per year?

A. £3.68m B. £6.82m C. £8.64m D. £10.8m E. £11.4m

MOCK PAPER B SECTION 1A

Questions **12** and **13** refer to the following passage:

- In the year ending June 2013 there were 1,730 fatalities in reported personal injury accidents, a 3 per cent drop from 1,785 in the year ending June 2012. The number of killed or seriously injured (KSI) casualties fell by 5 per cent, to 23,530, and the total number of casualties fell by 7 per cent to 188,540.
- A total of 8,560 car users were reported killed or seriously injured in the year ending June 2013, a fall of 6 per cent from the previous 12-month period.
- KSI casualties for the vulnerable road user groups – pedestrians, pedal cyclists and motorcyclists – showed overall decreases of 7, 1 and 6 per cent respectively compared with the year ending June 2012.
- The casualty rate per billion vehicle miles decreased for all casualty severities in the year ending June 2013, with falls of 3 per cent for fatalities, 6 per cent for serious injuries and 7 per cent for all casualties. This is the first publication in which the Department has included quarterly casualty rates.
- There were also significant decreases in the number of child casualties (aged 0-15) which fell from 18,166 in the year ending June 2012, to 15,920 in the year ending June 2013, a fall of 12 per cent. The number of child KSIs also fell in the same period by 11 per cent to 2,080. The number of child pedestrian casualties who were killed or seriously injured fell by 8 per cent to 1,440 in the year ending June 2013.
- There were drops in the number of accidents on all road types in the year ending June 2013 relative to the year ending June 2012. The number of fatal or serious accidents fell by 7 per cent on major roads (motorways and A roads) and 4 per cent on minor roads. On roads with speed limits over 40 mph (non-built up) fatal and serious accidents fell by 6 per cent and on roads with speeds limits up to an including 40 mph (built-up) they fell by 5 per cent.
- There were 185,540 casualties from 139,350 accidents in the year ending June 2013 which represents a 6 per cent fall for accidents and a 7 per cent fall for casualties compared with the year ending June 2012.

Question 12

Regarding the passage, which of these statements can be known to be true?

A. Child casualties are on the rise

B. Annual road deaths in the UK are falling

C. Vulnerable road users are more likely to be injured per vehicle mile than drivers

D. From June 2012 to June 2013, there were 188,540 serious injuries

E. Motorways are safer than built-up roads

Question 13

"The government is always under pressure to reduce road casualties. For this reason, anti-drink-drive campaigns costing millions of pounds are commonly produced, particularly around Christmas time. To address a one-year increase in drink driving related deaths, a new campaign was introduced. Subsequently, drink-driving casualties fell. The government therefore concluded that the £8m campaign had been a success."

Which of the following most undermines this argument?

A. Fewer people drink-drive these days than 10 years ago

B. Correlation does not imply causation: there is no plausible mechanism for the campaign to provide benefit.

C. The effect is too rapid for this campaign to have changed the public's attitude

D. Regression to the mean explains this phenomenon: values which were abnormally high one year are likely to settle down the next

E. When spending so much money, benefits are certain. The true test is in running a smaller campaign.

Question 14

"Some people with a sore throat and a chest infection have the 'flu."

Which of the following statements is supported?

A. Some people have a chest infection, but do not have the 'flu

B. Some people with a sore throat and a chest infection do not have the 'flu

C. Kate has the 'flu. Therefore she has a sore throat

D. The 'flu is defined as a sore throat and chest infection together

E. None of the above

Question 15

Catherine has 6 pairs of red socks, 6 pairs of blue socks and 6 pairs of grey socks in her drawer. Unfortunately, they are not paired together. The light in her room is broken so she cannot see what colour the socks are. She decides to keep taking socks from the drawer until she has a matching pair.

What is the minimum number of socks she needs to take from the drawer to guarantee at least one matching pair can be made?

A. 2 B. 3 C. 4 D. 5 E. 6

Question 16

Luca and Giovanni are waiters. One month, Luca worked 100 hours at normal pay and 20 hours at overtime pay. Giovanni worked 80 hours at normal pay and 60 hours at overtime pay. Neither received any tips. Luca earned €2000; Giovanni earned €2700. What is the overtime rate of pay?

A. €10 per hour B. €15 per hour C. €20 per hour D. €25 per hour E. €30 per hour

Question 17

"Train A leaves Plymouth at 10:00 and travels at 90mph. Train B leaves Manchester at 10:30 and travels at 70mph. The distance between the two cities is 387.5 miles. Due to a mistake, both trains are travelling on the same track." Calculate the distance from Plymouth at which the trains will collide.

A. 158 miles B. 203 miles C. 228 miles D. 248 miles E. 263 miles

Question 18

"100 pieces of rabbit food will feed one pregnant rabbit and two normal rabbits for a day. 175 pieces of food will feed two pregnant and three normal rabbits for a day. There is no excess food."

Which statement is **NOT** true?

A. A normal rabbit can be fed for longer than a day with 30 pieces of food.
B. 70 pieces of food are sufficient to feed a pregnant rabbit for a day.
C. A pregnant rabbit needs twice as many pieces per day as a normal rabbit.
D. Two pregnant and four normal rabbits will need 200 pieces of food for a day.
E. Three pregnant and ten normal rabbits will need 450 pieces of food for a day.

Question 19

"Studies of the brains of London taxi drivers show that training for "the knowledge", a difficult exam requiring knowledge of 20,000 London streets, enlarged a part of the brain believed to be important for spatial and organisational memory. This shows the brain can adapt to training and increase its abilities. Therefore if I wanted to improve my ability to remember names, I should also train my brain with repetitive tasks."

Which of the following **best** represents the flaw in this argument?

A. Enlarging of the brain does not necessarily mean it has improved
B. It might not be true to assume name memory and spatial memory use the same part of the brain
C. The brain enlargement would likely have happened anyway even without training
D. We do not know how London taxi drivers prepare for "the knowledge"
E. Practice does not necessarily improve performance on memory tasks

Question 20

"Michael bought a painting at an auction for £60. After 6 months, he realised the value of the painting had increased, so he sold it for £90. Realising a mistake, he wanted to buy the painting back, which he was able to do for £110. A year later, he then re-sold the painting for £130."

What is the total profit on the painting?

A. £20 B. £30 C. £40 D. £50 E. £60

Question 21

"Insect pests such as aphids and weevils can be a problem for farmers, as they feed on crops, causing destruction. Thus, many farmers spray their crops with pesticides to kill these insects, increasing their crop yield. However, there are also predatory insects such as wasps and beetles that naturally prey on these pests – which are also killed by pesticides. Therefore, it would be better to let these natural predators control the pests, rather than by spraying needless chemicals."

Which of the following best describes the flaw in this logic?

A. Many pesticides are expensive, so should not be used unless necessary
B. It fails to consider other problems the pesticides may cause
C. It does not explain why weevils are a problem
D. It fails to assess the effectiveness of natural predators compared to pesticides
E. It does not consider the benefits of using fewer pesticides

Question 22

A parliament contains 400 members. Last election, there was a majority of 43% of the popular vote to the liberal party. However, as a first-past-the-post system of constituencies was in effect, they gained 298 seats in parliament. How many excess members did they have, relative to a straight proportional representation system?

A. 72   B 98   C 112   D 126   E 148

END OF SECTION

Section 1B

Question 23

Consider the infinite series, $x - \left(\frac{1}{2}\right)x^2 + \left(\frac{1}{4}\right)x^3 - \left(\frac{1}{8}\right)x^4 \ldots$ Given that we know that the fifth term of the series is $\left(\frac{1}{32}\right)$, what is summation of the series given that the series converges as it heads toward infinity?

A $\dfrac{16^{\frac{1}{5}}}{2 + \frac{(16^{\frac{1}{5}})}{2}}$

B $\dfrac{1}{1 - (32)^{\frac{1}{4}}}$

C $\dfrac{8^{\frac{1}{5}}}{1 + 8^{\frac{1}{5}}}$

D $\dfrac{2}{2 - (16)^{\frac{1}{4}}}$

E $\dfrac{-2}{2 + (16)^{\frac{1}{4}}}$

F $\dfrac{1}{64 - 8^{\frac{1}{5}}}$

Question 24

If $\log_2 3 . \log_3 4 . \log_4 5 \ldots \log_n(n + 1) \leq 10$, what is the largest value of n that satisfies this equation?

A. 1022 B. 824 C. 842 D. 1023 E. 1020 F. 890

Question 25

a,b,c is a geometric progression where a,b,c are real numbers. If $a + b + c = 26$ and $a^2 + b^2 + c^2 = 364$, find b.

A. $\dfrac{1}{3\sqrt{25}}$ B. 6 C. $2\sqrt{6}$ D. 9 E. 4 F. $2\sqrt{3}$

Question 26

Given that a>0, find the value of a for which the minimal value of the function $f(x) = \left(a^2 + 1\right)x^2 - 2ax + 10$ in the interval $x \in [0; 12]$ is $\dfrac{451}{50}$.

A. 7 B. 12 C. 5 D. $\dfrac{50}{125}$ E. 8 F. 10

Question 27

If the probability that it will rain tomorrow is $\frac{2}{3}$ and the probability that it will rain and snow the following day is $\frac{1}{5}$, given that the probability of rain and snow occurring on any given day are independent from one another, what is the probability that it will snow the day after tomorrow?

A. $\dfrac{10}{3}$ B. $\dfrac{3}{10}$ C. $\dfrac{2}{15}$ D. $\dfrac{15}{2}$ E. $\dfrac{4}{9}$ F. $\dfrac{1}{5}$

MOCK PAPER B SECTION 1B

Question 28

If $cos\,2\theta = \frac{3}{4}$, then $\frac{1}{cos^2\theta - sin^2\theta} =$

A. $\frac{4}{3}$ B. 4 C. -1 D. $\frac{3}{4}$ E. 2 F. 1

Question 29

Describe the geometrical transformation that maps the graph of $y = 0.2^x$ onto the graph of $y = 5^x$.

A. Reflection in the x-axis

B. Reflection in the y-axis

C. Multiplication by a scale factor of 25

D. Addition of the constant term 4.8

E. Multiplication by scale factor of 5

F. Multiplication by scale factor $\frac{1}{25}$

Question 30

Find the solution to the equation $log_4(2x + 3) + log_4(2x + 15) - 1 = log_4(14x + 5)$

A. There is no solution B. $\frac{2}{5}$ C. $\frac{5}{2}$ D. -1 E. 1 F. 0

Question 31

The normal to the curve $y = e^{2x-5}$ at the point $P(2, e^{-1})$ intersects the x-axis at the point A and the y-axis at the point B. Which of the following is an appropriate formula for the area of the triangle that is formed in terms of e, m, and n, where m and n are integers?

A. $\frac{(e^2+1)^m}{e^n}$

B. $\frac{(e^3+1)^{\frac{1}{n}}}{m}$

C. $\frac{e^n}{(e^2+1)^m}$

D $\frac{m^{\frac{1}{n}}}{e^3+1}$

E $\frac{e^{2m}}{e^n+1}$

F $\frac{(e^2-1)^m}{e^{2n}}$

Question 32

Given that $secx - tanx = -5$, find the value of cos x.

A. -0.2 B. 0.2 C. $\frac{-13}{5}$ D. $\frac{-5}{13}$ E. 0.5 F. -0.5

Question 33

Consider the line with equation $y = 2x + k$where k is a constant, and the curve $y = x^2 + (3k - 4)x + 13$. Given that the line and the curve do not intersect, what are the possible values of k?

A. $\frac{-1}{3} < k < 3$ B. $\frac{-4}{9} < k < 4$ C. $\frac{1}{2} < k < \frac{5}{3}$

D $\frac{3}{2} < k \leq \frac{8}{3}$ E $\frac{1}{3} < k < 3$ F $-3 < k < \frac{1}{3}$

Question 34

A circle with centre C(5,-3) passes through A(-2,1), and the point T lies on the tangent to the circle such that AT = 4. What is the length of the line CT?

A. 9 B. 18 C. $\sqrt{95}$ D. $8\sqrt{2}$ E. $\sqrt{69}$ F. 8

Question 35

Evaluate: $(6\sin x)(3\sin x) - (9\cos x)(-2\cos x)$

A. 0 B. 0.5 C. 1 D. -1 E. 18 F. -18

Question 36

In the figure to the right, all triangles are equilateral. What is the shaded area of the figure in terms of r?

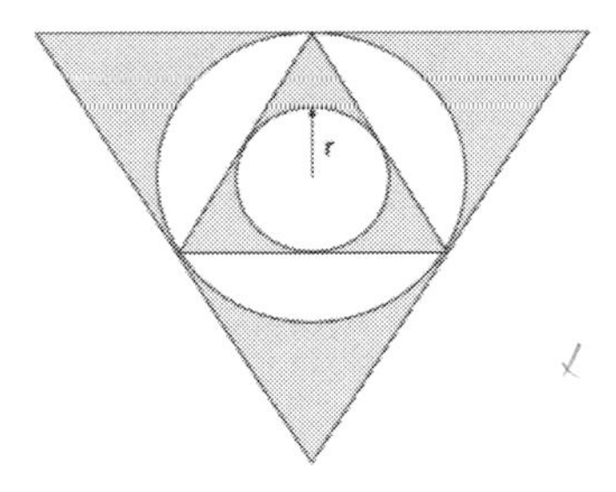

A. $5r^2\left(2\sqrt{6} - 3\pi\right)$

B. $5r^2\left(5\sqrt{2} - 6\pi\right)$

C. $5r^2\left(3\sqrt{3} - \pi\right)$

D. $5r^2\left(4\sqrt{3} - 2\pi\right)$

E. $5r(2\sqrt{6} - 3\pi)$

F. $5r^2\left(5\sqrt{2} + 6\pi\right)$

Question 37

Suppose I use a binomial expansion to determine the value of $(3.12)^5$. What is the minimum number of terms that I must obtain in the expansion of $(3.12)^5$ in order to receive a result accurate to 1 decimal place?

A. 4 B. 5 C. 6 D. 7 E. 9 F. 8

MOCK PAPER B SECTION 1B

Section 2

Read the extract taken from Simon Tait's *Maria Miller: thank you and goodbye_* (2014, *The Stage*- edited) and then answer the question below in the space provided in this booklet.

Your answer will be assessed taking into account your ability to construct a reasoned, insightful and logically consistent argument with clarity and precision.

To what extent does art and culture merit public funding?

Maria Miller: thank you and goodbye

She "has done an effective job in making the case for the value of public funding" Bazalgette says in response to the news that the arts are to get a ring-fenced 5% cut when ACE had been told to model for 10% and 15% scenarios for 2015/16.

But it is rather damning with faint praise. He goes on to say: "It is hugely encouraging that the Chancellor and the Treasury have listened to the argument that the arts and culture makes such a valuable contribution to our quality of life and the economy".

The argument made by the arts and culture, note, not the secretary of state.

DCMS (Department for Culture, Media and Sport) as a whole has got a less generous cut of 8%, and the deal for the arts appears to have been negotiated separately by the likes of Bazalgette and national museum directors like Nicholas Serota who, three weeks ago, went to George Osborne directly with economic arguments for a more lenient treatment of the sector. It was at this point that Osborne and the Treasury finally "got it" and realised how damaging a bigger arts cut would be to the economy for negligible saving.

It means that Mrs Miller cannot simply pass on to the arts the 8% cut as she and her predecessor, Jeremy Hunt, have done in the past because there is no fat in the DCMS, having been cut to the bone already, to absorb a new reduction itself. As it is, she will have to find the saving from elsewhere in her budget.

Nevertheless, she has hung out and got a better settlement than most other government departments who are suffering at least 10% reductions as the government tries to find more savings, but it seems the knives are out not for culture or the arts but for Miller herself.

The knives appear to be out for her in government for a number of reasons, including not dealing decisively with Leveson. The culture secretary has also been the subject of unprecedented vilification in the Tory press, with the Daily Mail's drama critic Quentin Letts declaring a couple of weeks ago that "Culture is the department where a country can assert its character. If only its Secretary of State had one". In May, she made her first speech on the arts, calling for the economic argument to be made, Letts conceded, but "Where was the question of morality in Mrs Miller's approach to the arts? Where was the vision that the arts can civilise us? Where was an idea of the arts as the most meritocratic of gifts, a route which can offer talented and aspiring youngsters a route to self-fulfilment…? There is not even much impression she is an arts lover. It was a speech that could have been given by any one of her departmental officials".

Her desperate attempts to grab a positive headline culminated last week in a damp squib of an announcement about the First World War centenary commemoration, in which nothing new was announced (except that 600-odd streets in England were to be renamed after VC winners from the Great War), and the major news about the cultural element cannot be revealed before August. On Friday, The Times's normally gentle columnist Richard Morrison wrote that "Some (culture secretaries) have been bores; some bluffers. But not one has depressed me as Maria Miller does.

As for the arts, the triumph is substantial and this might be a seachange in the way governments see the sector. The Arts Council, as fuel for the Bonfire of the Quangos, has taken an enormous battering since 2010 and the sector has correctly acknowledged that there is no case for "special treatment" while cuts amounting to 33% have been meted out, and of 50% to ACE itself. But now culture has established the principle that it is a special case after all, and with sense and imagination much of the effect of the new 5% cut might be ameliorated through the National Lottery.

The question now is whether that principle will be accepted by the other great subsidisers of the arts, the local authorities in whose hands the futures of dozens of theatres lie and whose extreme economic pain is even greater than Osborne's.

END OF PAPER

ANSWERS

Answer Key

Paper A	Paper B
1 B	**1** C
2 D	**2** C
3 E	**3** B
4 C	**4** D
5 C	**5** D
6 C	**6** E
7 E	**7** B
8 C	**8** C
9 D	**9** C
10 D	**10** D
11 C	**11** C
12 C	**12** B
13 D	**13** D
14 B	**14** E
15 B	**15** C
16 E	**16** D
17 C	**17** D
18 C	**18** E
19 C	**19** B
20 B	**20** D
21 C	**21** D
22 E	**22** D
23 B	**23** A
24 B	**24** D
25 C	**25** B
26 D	**26** A
27 A	**27** B
28 B	**28** A
29 B	**29** B
30 B	**30** C
31 B	**31** A

MOCK PAPER B SECTION 2

32 A	**32** D
33 C	**33** B
34 D	**34** A
35 D	**35** E
36 A	**36** C
37 D	**37** A

MOCK PAPER B SECTION 2

Mock Paper A Answers

Section 1A

Question 1: B
James runs 26.2 seconds, which is outside the qualifying time, therefore he does not qualify

Question 2: D

5.6/7 gives the unit price of 80p – this equals a packet of crisps. Multiplying this by 2 gives the sandwich and by 4 gives the watermelon price of £3.20

Question 3: E

Jane leaves at 2:35pm and arrives at 3:25pm, taking 50 minutes. Sam's journey takes twice as long, so leaving at 3:00pm it takes 100 minutes, giving an arrival time of 4:40pm

Question 4: C

After the donation, Sam has eight sweets. Therefore Hannah had 16 sweets after the transaction and hence 13 sweets before

Question 5: C

Find original pay: £250/0.86 = 290 basic original pay. Add the rise: (290 x 1.05) + 6 = £311 new basic pay. Subtract the income tax at 12% = 311 x 0.88 = £273 new pay rate

Question 6: C

Given the first cube is a white cube, you are drawing from one of three boxes, boxes A, C or D. Boxes C and D will have just had their only white cube removed, whereas box A will have one white cube remaining. Therefore the probability of drawing a second white cube is $^1/_3$, thus the probability of non-white (i.e. black) is $^2/_3$.

Question 7: E

This is a simultaneous equations question. $500 + 10(x - 80) = 600 + 5x$; true when $x \geq 80$.
$500 + 10x - 800 = 600 + 5x$
» $5x = 900$
» $x = 180$

Question 8: C
If eating more slowly caused a reduction in the time available to work, the candidate might be less productive.

Question 9: D

This is a LCM question. We need to find the lowest common multiple of the song lengths. The LCM of 100, 180 and 240 is 3,600 seconds – equal to 60 minutes. For ease of arithmetic, you may choose to work reduce all numbers by a factor of 10.

Question 10: D

The journey is 3 hours and 45 mins, minus a 14 minute break gives 3hrs 31 mins travel time, or 211 minutes. Therefore the average speed is 51mph, or 82kmh by using the stated conversion factor.

Question 11: C

The mean guess is £13.80, which is £5.80 too high.

Question 12: C

The overall error for respondent 3 is £13, which is the least

Question 13: D

Scale back and forth from known quantities. Country B has 32.1m so Country D has 38.6m people.

Question 14: B

Country B has 32.1m people. Therefore \$45 x 32.1m = \$1.45bn

Question 15: B

The average speed is 24mph, independent of distance travelled as it cancels. Imagine this covers a set distance of say 30 miles. It will take 1 hour on the way and 1.5 hours on the way back. 60/2.5 = 24. This is true of all distances, the ratio is the same.

Question 16: E

None of the above can be reliably deduced from the passage alone

Question 17: C

Imagine the toothpaste costs 100p originally, and follow the price through. It rises by 80% to 180p, then is reduced by 50% to 90p. Three tubes are purchased for the price of 2 (i.e. 180p), therefore the cost per unit is 180/3 = 60p. 60p = 60% x 100, the original price

Question 18: C

Argument C has the same form, asserting that since something is not happening, the result of the action will never be true.

Question 19: C

Statement C is the only one making reference to the potential outcome of solving crimes faster, thereby providing a plausible mechanism for a reduction in cybercrime rates

Question 20: B

The passage suggests that the attacks were carried out by extra terrestrial beings. Though the supposed UFO sightings have rational explanations, the writer feels this is insufficient to dismiss his idea.

Question 21: C

The initial argument suggests that two things must be present for an action to happen. If only one is absent, the action cannot happen. Argument C has the same form, the others do not.

Question 22: E

Growing vegetables needs several positive traits. The passage does not tell us which is the most important or most commonly lacked skill, only that more than one skill is required for success.

END OF SECTION

Section 1B

Question 23: B
We know that the product of slopes of perpendicular lines equals -1.
Therefore:
$(n+1)(n+3) = -1.$
$n^2 + 4n + 3 = -1$
$n^2 + 4n + 4 = 0$
Factorising gives (n+2)(n+2), therefore n = -2 for the lines to be perpendicular.

Question 24: B
Algebraically, we can find the result of reflecting the curve $y = x^2 + 3$ across the line y=x by replacing y with x in the equation, and solving for the value of y in order to find the relevant equation, which is:

$x = f(y) = \sqrt{y-3}$
Replacing y with x gives:
$y = \sqrt{x-3}$

Translating the resulting equation by $\begin{pmatrix} 4 \\ 2 \end{pmatrix}$ corresponds to introducing (-4) to the x term and (+2) to the y:

$y + 2 = \sqrt{x-4-3}$
$y = \sqrt{x-7} + 2$
The x-intercept is found by setting f(x) = 0.
$\sqrt{x-7} + 2 = 0$
$\sqrt{x-7} = -2$
$x - 7 = 4$
$x = 11$

Question 25: C
Take logs of each side and separate out the LHS:
$3x\log_{10}a + x\log_{10}b + 4x\log_{10}c = \log_{10}2$
$x(3\log_{10}a + \log_{10}b + 4\log_{10}c) = \log_{10}2$
$x\log_{10}\left(a^3bc^4\right) = \log_{10}2$
$$x = \frac{\log_{10}2}{\log_{10}\left(a^3bc^4\right)}$$

Question 26: D
$Let\ y = 2^x.\ Then,\ y^2 - 8y + 15 = 0.$
Solving this either using the quadratic equation or otherwise, we obtain y = 3 or y = 5.

If $3 = 2^x \rightarrow x = log_2 3 = \frac{log_{10}3}{log_{10}2}$

If $5 = 2^x \rightarrow x = \frac{log_{10}5}{log_{10}2}$.

The sum of the roots is $\frac{log_{10}3}{log_{10}2} + \frac{log_{10}5}{log_{10}2} = \frac{log_{10}(3*5)}{log_{10}2} = \frac{log_{10}15}{log_{10}2}$

Question 27: A
Recall the discriminant condition for the existence of real and distinct roots,
$b^2 - 4ac > 0$
Using the coefficients in our question, this is: $(a-2)^2 > 4a(-2)$
$a^2 + 4a + 4 > 0$

$(a+2)^2 > 0$
Since this is a squared number, all values but a = -2 will satisfy this equation.

Question 28: B

We can use the inclusion-exclusion principle to find the probability that none of the balls are red.
Since there are 2n blue balls, n red balls, and 3n balls altogether, the probability of drawing no red balls within the two draws is: $\frac{2n}{3n} \times \frac{(2n-1)}{(3n-1)} = \frac{4n-2}{3(3n-1)}$

Therefore, the probability of drawing at least one red ball is equal to:
$$1 - \frac{4n-2}{3(3n-1)} = \frac{3(3n-1)-(4n-2)}{3(3n-1)} = \frac{9n-3-4n+2}{3(3n-1)} = \frac{5n-1}{3(3n-1)}$$

Question 29: B

The numerator of $\frac{x^2-16}{x^2-4x}$ is in the form $a^2 - b^2$, which means that it can be expressed as the quantity $(a+b)(a-b) = (x+4)(x-4)$
In turn, the numerator can be simplified into: $x(x-4)$.
$\frac{x^2-16}{x^2-4x}$ can therefore be expressed as: $\frac{(x+4)(x-4)}{x(x-4)}$

Which simplifies to: $\frac{(x+4)}{x}$

Question 30: B

At the stationary point, $\frac{dy}{dx} = 0$. Using the product rule: $\frac{dy}{dx} = x^2e^x + e^x \times 2x$

When $\frac{dy}{dx} = 0$, $x^2e^x + e^x \times 2x = 0$

Hence, $xe^x(x+2) = 0$
Which shows that the x-coordinates passing through the stationary points of $y = x^2e^x$ are x=0 and x= -2 respectively.
Therefore, the equation of the quadratic function is: $x(x+2) = x^2 + 2x$.

Question 31: B

First, we set the two equations equal to one another: $k(x+4) = 8 - 4x - 2x^2$
$2x^2 + kx + 4x + 4k - 8 = 0$
$2x^2 + (k+4)x + 4(k-2) = 0$
Subsequently, we set $b^2 - 4ac = 0$, as follows: $(k+4)^2 - 4 \times 2 \times 4(k-2) = 0$
$k^2 - 24k + 80 = 0$
Solving this equation yields: $k = 4$, $k = 20$

Question 32: A

The expansion of $y_1 = (1-x)^6 = 1 - 6x + 15x^2$
The expansion of $y_2 = (1+2x)^6 = 1 + 12x + 60x^2$
The ratio of the second coefficient of y_1 to the third coefficient of y_2 is $-\frac{6}{60} = -\frac{1}{10}$.

Question 33: C

These integers form an arithmetic progression with 300 terms, where n = 300, $a_1 = 1$, and a_n= 300. If you substitute these values into the formula for the sum of a finite arithmetic sequence, you will get:

$S_n = 1 + 2 + 3 + 4 + 5 + \ldots + 300$
$S_n = \frac{n}{2}(a_1 + a_n)$
$S_n = \frac{300}{2}(1 + 300)$

$S_n = 150(301) = 45150$

Question 34: D

Recall the double angle formula for sine: $\sin 2\theta = 2\sin\theta\cos\theta$

Since $\sin 2\theta = \frac{2}{5}$, $2\sin\theta\cos\theta = \frac{2}{5}$, $\sin\theta\cos\theta = \frac{1}{5}$

$$\frac{1}{\sin\theta\cos\theta} = \left(\frac{1}{\frac{1}{5}}\right) = 5$$

Question 35: D

$-11 + 4\lceil n \rceil = 5$

$4\lceil n \rceil = 16$

$\lceil n \rceil = 4$

Since 4 is the greatest integer less than or equal to n, n must be on the interval $4 \leq n < 5$.

Question 36: A

If $a \varnothing b = \frac{a/2}{b}$, then $10 \varnothing n = \frac{10/2}{n} = \frac{5}{n}$ - Equation 1

$n \varnothing \frac{1}{10} = \frac{(n/2)}{\frac{1}{10}} = \frac{5n}{1} = 5n$ - Equation 2

Setting Equation 1 and Equation 2 equal to one another, $5n = \frac{5}{n}$

Thus, $5n^2 = 5$, $n^2 = 1$, and $n = \pm 1$.

Question 37: D

Since -1 is a zero of the function, $(x + 1)$ is a factor of the overall polynomial. By long division or synthetic division, we can determine that $\frac{2x^3 + 3x^2 - 20x - 21}{x + 1} = 2x^2 + x - 21$.

Factoring $2x^2 + x - 21 = 0$, we get: $(2x + 7)(x - 3) = 0$

The roots are $x = -\frac{7}{2}$ or $x = 3$.

END OF SECTION.

Section 2

Introduction:

- Define "value". Provide a provisional definition, describing value in terms of the benefit that is generated for consumers by a particular product. In this passage, we described the phenomenon of 'price framing', which is the practice of setting prices such that the consumer perceives something as valuable as a result of the price that is offered, although this value need not directly correspond to an objective dollar valuation, as we have seen in the case of the crafty seller.
- The key question: To what extent does price reflect value to a consumer, should perception of value in and of itself be taken as value in and of itself? Who is qualified to speak of 'value' and how can this be regulated, if at all?

Paragraph 1:

- You can suggest that pricing reflects value insofar as it corresponds to a dollar sacrifice that is made in order to obtain a particular product. Prices in a free market are set by the interactions of buyers and sellers, who set prices on the basis of their business requirements – Insofar as we are consumers, we do not actually know what a business's private valuation of the good they are selling is, and therefore we can leave reason to things that reason deserves. Even if we were to take issue with a particular discount, this is the word of the consumer against the word of the seller, and the seller privately knows the price that they wanted to sell at, unless a specific rule for pricing was determined beforehand, it was known, and it was clear that it was deviated from.

<u>- Passage Example:</u>

Price may reflect value insofar as it corresponds to a dollar sacrifice that a consumer makes in order to purchase or obtain a product. Economic theory suggests that if the price were set wrongly, then people would simply not pay. On some level, consumers 'get what they pay for', which suggests that things are priced fairly, but it depends on how you define 'value', because no matter what the promotion was set at, if the consumer paid, then there is no problem.

Suppose we define value as what consumers are willing to pay. If we abide by what we observe from consumer behaviour in response to misleading pricing, there are two possible explanations, one of which is that consumers are irrational and unable to distinguish value accurately or with certainty save for through heuristics and hence are easily deceived, and the second of which is that consumers place a valuation not just on the product in and of itself, but on the nature of price cut displays or the limited time promotion, which can provide the psychological reassurance that the consumer is getting a good deal or affect how much they value the deal, which is priced into their decision when considering different alternatives.

In this line of argument, perception equals value, and the good is priced correctly. Government need not do anything.

Paragraph 2:

- You can suggest that price can be used deceptively, as in the examples that have been given in the text. It is completely possible that businesses can use prices in order to deceive, mislead, to suggest that something is worth more than it actually is, or to imply that something has a value that it never had – But subsequently note that that is based on a specific definition or designation of value, which may not in fact have an objective basis.

- Passage Example:

Suppose that we define value in absolute dollar terms, and suggest therefore that there is an absolute valuation for a good that consumers misperceive. Behavioural economics suggests that humans are not completely rational in determining price, and pricing strategies play into the systematic biases that they have, suggesting that humans can systematically misperceive the absolute value of a product if they fail to consider their biases. For example, suppose the example of a single product that is marketed differently, one presenting the benefits of the good alone, and the other presenting the potential loss that a customer might sustain if they did not purchase that good. Kahneman and Tversky write that human beings exhibit loss aversion, which suggests that if they had the opportunity to pass up what has been presented as a good discount, they may consider it to be more favourable relative to the scenario in which they had been presented with a good deal, even though there was a true absolute valuation, and the consumer was wrong on every count.

- Point out that the role of government, if viewed as protecting consumers, suggests that if this view is held, then the government should regulate.

Paragraph 3:

- To what extent regulating price-setting makes sense, and some possible regulatory schemes.

- Passage Example:

How governments treat 'value' affects how they in turn treat the question of whether they should regulate at all, that is, if you even consider the free decisions of consumers to purchase and sellers to set prices to be a domain in which the government should interfere in the first place. Supposing you do, then your decision may be moderated by whether you believe value is solely determined by perception of the consumer, or there is an objective price for a specific good that should be arrived at independently by different individuals, free markets should theoretically allow for people to set price in accordance to their desires, and to have businesses live or die depending on whether they set the price too high and subsequently receive no customers, or set it too low.

- Make some possible arguments for regulating the information that sellers provide in their prices, and outline their implications. What would happen if we implemented a 'no reference pricing' policy? Would outcomes be fairer, less fair, would they be better? According to what dimension? If 'value' is simply defined as what consumers are willing to pay in dollars, what is better and what is worse? If it is defined as something that is absolute but that consumers routinely misperceive, what is the implication?

Conclusion

Summarize the main points:

- How you treat 'value' affects the particular way you consider a particular price.
- You may consider 'value' to reflect just the perception of the consumer, in which case there may not be a problem.
- On the other hand, you may consider there to be an objective 'value' that consumers misperceive, in which case then perhaps there is an issue.
- How you treat value affects how you will regulate, if at all you believe that regulation should be implemented.
- Zoom out and say why the question really matters.
- Price reflects value, but whether this value is solely contingent on perception or showcases something absolute, is something that is up in the air.
- This should affect our decisions and views concerning whether to regulate or not to regulate accordingly.

Mock Paper B ANSWERS

Section 1A

Question 1: C

Joseph does not have blue cubic blocks, since all his blue block are cylindrical.

Question 2: C

130°. Each hour is 1/12 of a complete turn, equalling 30°. The smaller angle between 4 and 8 on the clock face is 4 gaps, therefore 120°. In addition, there is 1/3 of the distance between 3 and 4 still to turn, so an additional 10° must be added on to account for that.

Question 3: B

The chance of red is 2/6 = 1/3. To get no reds at all, it must be non-red for each of three independent rolls. The probability of this is $(2/3)^3$ = 8/27. Therefore the probability of at least one red is 1 – 8/27 = <u>19/27</u>

Question 4: D

These three furniture items are compatible with having 6 legs. All the other statements are false.

Question 5: D

Work this out by time. The friends are closing on each other at a total of 6mph overall, therefore the 42 miles take 7 hours. In seven hours, the pigeon, flying at 18moh covers 18 x 7 = 126 miles.

Question 6: E

The passage does not make any supported claims about fruit juice. It gives rationale for both benefits and risks of fruit juice consumption without reaching a conclusion.

Question 7: B

Calculate the overall cost of three stationery sets, then subtract any items not bought. For each item shared between two people, there is one of that item not required. The overall cost is £6.00 per person, £18.00 overall. Subtract one geometry set (£3), one paper pad (£1) and one pencil (50p) to give £13.50 overall cost.

Question 8: C

Argument C is the most convincing. It gives a strong rationale as to why the notion that people should only pay for services which they personally use is likely to have serious adverse consequences on the nation as a whole. Therefore this flawed logic is not suitable to apply to the arts funding dilemma.

Question 9: C

Moving matches 1 and 4 to form a cross inside one of the other cubes will solve the problem. Two squares are broken (the top left hand corner and the overall large square) but four new small ones are created, bringing the total up to seven.

Question 10: D

The white square is opposite brown, since both are adjacent to blue on opposite sides. White and brown cannot be adjacent to each other since the position of the opposite black and red sides makes that impossible.

Question 11: C

We take the overall price to the UK and subtract money which does not go to the farmers. 36,000,000kg at 300p/kg gives £108m. Subtract commission 108 x 0.8, then take 10% of the remaining proceeds as the farmers' share, giving £8.64m

Question 12: B

The first paragraph tells us annual road deaths have fallen, so B is true. The others are false.

Question 13: D

Regression to the mean is a phenomenon observed when a value is variable within a probability distribution. Sometimes by chance it will be at the high or low end, but thereafter it is likely to be closer to what is expected. This can explain the fall in drink driving deaths after the new campaign.

Question 14: E

None of the responses can be reliably deduced from the statement regarding the 'flu.

Question 15: C

Catherine must choose four socks. If choosing three or fewer, it is possible that they could each be of different colour. When choosing four, it is certain that at least two socks will make a matching pair, but possible that there will be two pairs.

Question 16: D

This is another simultaneous equations question. Solve to find x, the normal rate of pay.

$100x + 20y = 2000$ » $60y = 6000 - 300x$ (substitute this)

$80x + 60y = 2700$

» $80x + (6000 - 300x) = 2700$

» $220x = 3300$

$x = 15$

Question 17: D

This is demanding question - the easiest way to do this is via simultaneous equations. Let A be the distance travelled by the Plymouth train and B the distance travelled by the Manchester train. Thus:
$A = 90y + (90 \times \frac{3}{4})$ and $B = 70y$

The collision will occur when the total distance travelled by both trains is = 387.5

i.e. $A + B = 387.5$

Therefore, $90y + 67.5 + 70y = 387.5$

Solving gives $y = 320/160$

$y = 2$ hours. Thus, the collision happens 2 hours after Train B leaves at 10:45 i.e. at 12:45.

The distance of train A from Plymouth at 12:45 is given by:

$= 90 \times 2$ hours $+ 67.5 = 247.5$ miles, which rounds to 248 miles

Question 18: E

Statement E is not true, the others are true. A pregnant rabbit requires 50 pieces per day and a normal rabbit requires 25. Therefore three pregnant and ten normal rabbits require only 400 pieces per day, not 450.

Question 19: B

If memory of names uses a different part of the brain, then conclusions drawn from this experiment may have no validity.

Question 20: D

Michael pays £60 and £110 = £170 for the painting. He sells it for £90 and £130 = £220. Thus, he makes a profit of £220 - £170 = £50.

Question 21: D

The principle problem is that it does not compare the relative effectiveness of pesticides and natural predators. It might be that pesticides are far more effective at controlling pests, despite the unnecessary excess killing.

Question 22: D

Proportionately, there would be 172 members. Therefore there is an excess of 298 – 172 = 126 members.

END OF SECTION

Section 1B

Question 23: A

We can find the common ratio of the series by dividing the second term of the series by the first, yielding the common ratio $r = \left(-\frac{1}{2}\right)x$

Since we know that the fifth coefficient is equivalent to $\frac{1}{32}$, we can solve for the value of x, the first term in the series, by equating 1/32 to the formula for the fifth term of a geometric series:

$$\frac{1}{32} = ar^4$$

$$\frac{1}{32} = x(\left(-\frac{1}{2}\right)x)^4$$

$$\frac{1}{32} = \left(\frac{1}{16}\right)x^5$$

$$x^5 = \left(\frac{16}{32}\right)$$

$$x = \frac{(16)^{\left(\frac{1}{5}\right)}}{2}$$

This is an infinite geometric series with a first term of $a = x = \frac{(16)^{\left(\frac{1}{5}\right)}}{2}$. We can simply find the common ratio by substituting $r = \left(-\frac{1}{2}\right)x = \left(-\frac{1}{2}\right)\frac{(16)^{\left(\frac{1}{5}\right)}}{2}$.

The sum to infinity of a geometric series is given by $S_\infty = \frac{a}{1-r}$. Therefore, the sum of the series is given by:

$$S_\infty = \frac{\left(\frac{16^{\frac{1}{5}}}{2}\right)}{1-(-\frac{1}{2})\left(\frac{(16)^{\left(\frac{1}{5}\right)}}{2}\right)}$$

$$S_\infty = \frac{16^{\frac{1}{5}}}{2+\frac{(16^{\frac{1}{5}})}{2}}$$

Question 24: D

$$\log_2 3 \times \frac{\log_2 4}{\log_2 3} \times \frac{\log_2 5}{\log_2 4} \ldots \frac{\log_2(n+1)}{\log_2 n} \leq 10$$

Solving the above equation, we have that $\log_2(n+1) \leq 10$. Consequently, $n+1 \leq 1024$. The largest value of n that satisfies this equation is 1023.

Question 25: B

We have:

$$(a+b+c)^2 = a^2+b^2+c^2+2(ab+bc+ca) = 364+2(ab+bc+ca) = 26^2 = 676$$

so $ab+bc+ca = 156$.

Since b and c are the second and third terms of a geometric progression respectively, let us denote $b = ar, \ and \ c = ar^2$

We have $a + b + c = a + ar + ar^2 = 26$ and $ab + bc + ca = a^2r + a^2r^3 + a^2r^2 = 156$
$a(1 + r + r^2) = 26$ and $a^2r(1 + r + r^2) = 156 = 6 \cdot 26$.
We can divide both equations to get
$a^2r(1 + r + r2)/a(1 + r + r^2) = 6$, or $ar = b = 6$.

Question 26: A

$f(x)$ is a parabola, which is opened up (since its leading coefficient is $a^2+1>0$), so it has only one extremum and it is a global minimum. $f'(x)=0 <=> 2\left(a^2+1\right)x-2a=0,\ or\ x=\frac{a}{a^2+1}$. Luckily for us, $\frac{a}{a^2+1}=\frac{1}{2}\times\frac{2a}{a^2+1}\leq 1/2$

(since $0\leq\frac{2a}{a^2+1}\leq 1$ for any positive a).

As a result, the minimum in the interval is reached for $x=\frac{a}{a^2+1}$.

We substitute into $f(x)$ to reach

$$fmin(x)=f\left(\frac{a}{a^2+1}\right)=\left(a^2+1\right).\left(\frac{a}{a^2+1}\right)^2-2a\times\frac{a}{a^2+1}+10$$

$$=\frac{a^2}{a^2+1}-\frac{2a^2}{a^2+1}+10=10-\frac{a^2}{a^2+1}=\frac{9a^2+10}{a^2+1}$$

We want this value to be equal to $\frac{451}{50}$.

$\frac{9a^2+10}{a^2+1}=\frac{451}{50}$, so we cross multiply: $450a^2+500=451a^2+451,\ or\ a^2=49.$

Which means that a=7, since $a>0$.

Question 27: B

We know that rain and snow are independent events. If the probability that it will rain is $\frac{2}{3}$ and the probability that it will both rain and snow the following day is $\frac{1}{5}$, we can find the probability that it will snow the day after tomorrow by simply solving the equation.

$$\frac{2}{3}x=\frac{1}{5}$$

Which yields:

$$x=\frac{3}{10}$$

Question 28: A

Let us use the double angle formula, $\cos 2\theta = cos^2\theta - sin^2\theta$.

Given we know that $\cos 2\theta=\frac{3}{4}=cos^2\theta-sin^2\theta$, we know that $\frac{1}{cos^2\theta-sin^2\theta}=\frac{1}{\frac{3}{4}}=\frac{4}{3}$.

Question 29: B

If you draw the graphs, you will notice that the two graphs are the reflections of one another in the y-axis.

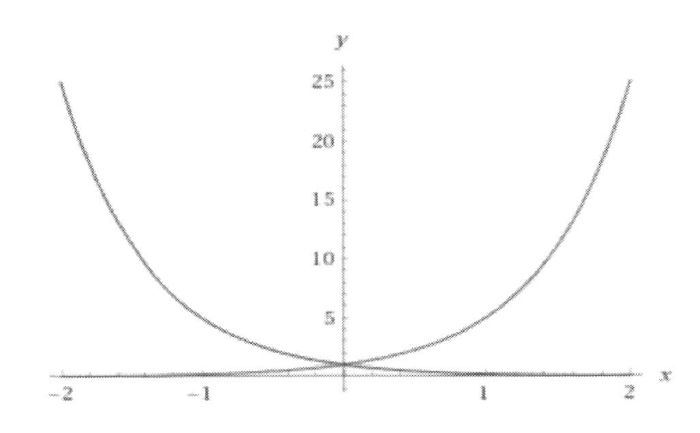

Question 30: C

Note that $1=\log_4(4)$.

$\log_4(2x+3)+\log_4(2x+15)-\log_4(4)=\log_4(14x+5)$

$\log_4(2x+3)(2x+15)=\log_4 4(14x+5)$

$(2x+3)(2x+15)=56x+20$

$4x^2+36x+45=56x+20$

$4x^2-20x+25=0$

By factoring,

$4x^2-20x+25=0$

$(2x-5)^2=0$

Hence, $x=\frac{5}{2}$

Question 31: A

The gradient of the curve is $\frac{dy}{dx} = 2e^{2x-5}$. We know that the gradient of the normal to the curve is $-\frac{1}{\frac{dy}{dx}}$.

Consequently, the equation of the normal is $y - e^{-1} = -\frac{e}{2}(x-2)$.

At the point A, where y=0, $x = 2 + \left(\frac{2}{e^2}\right)$

At point B, where x = 0, $\quad y = e + \frac{1}{e} = \frac{e^2+1}{e}$

Since the area of a triangle is $\frac{1}{2} \times Base \times Height$, the area of the triangle OAB is:

$$Area = \frac{1}{2} \times \frac{e^2+1}{e} \times 2 \times \frac{1+e^2}{e^2} = \frac{\left(e^2+1\right)^2}{e^3}$$

Question 32: D

We know that (sec x + tan x)(sec x – tan x) = $\sec^2 x - \tan^2 x$.

Using the trigonometric identity $\sec^2 x - \tan^2 x = 1$, as well as the information provided in the question, we know that:

$-5(\sec x + \tan x) = 1$

Therefore,

$(\sec x + \tan x) = -\frac{1}{5}$

By substitution, we know that $\sec x - \tan x + (\sec x + \tan x) = -5 + \left(-\frac{1}{5}\right)$

$2\sec x = -5.2$

$\sec x = -\frac{5.2}{2} = -2.6 = -\frac{13}{5}$

Since sec x = $\frac{1}{\cos x}$,

$\cos x = \frac{1}{\sec x} = -\frac{5}{13}$

Question 33: B

First, let us find the points along which any potential intersection between the line and the curve would take place, by setting the two equations equal to one another.

$x^2 + (3k-4)x + 13 = 2x + k$

$x^2 + 3kx - 6x + 13 - k = 0$

$x^2 + 3(k-2)x + 13 - k = 0$

Since the line and the curve do not intersect, we know that there must not be any real roots.

As such, by the discriminant condition, we know that $b^2 - 4ac < 0$.

Therefore:

$\left(3(k-2)\right)^2 - 4(13-k) < 0$

$9\left(k^2 - 4k + 4\right) - 52 + 4k < 0$

$9k^2 - 32k - 16 < 0$

$(9k+4)(k-4)$

We know that the critical values therefore extend from $-\frac{4}{9} < k < 4$.

MOCK PAPER B

Question 34: A

The distance AC (equivalent to the radius of the circle) can be determined given the coordinates of A and C:

A =(-2,1) C = (5,-3)

Therefore $AC = \sqrt{(5+2)^2 + (1+3)^2} = \sqrt{65}$

To find the length of the line CT, we use Pythagoras' Theorem:

$CT^2 = AT^2 + AC^2$

$CT^2 = 4^2 + 65$

$CT^2 = 81$

$CT = 9$

Question 35: E

$= (6\sin x)(3\sin x) - (9\cos x)(-2\cos x)$

$= 18\sin^2 x + 18\cos^2 x$

$= 18(\sin^2 x + \cos^2 x)$

$= 18$

Question 36: C

Define half the length of the inner equilateral triangle as x, and form a right-angled triangle by drawing a line from the centre of the inner circle to the inner triangle, defining the distance of that line as y.

$\tan 30 = \frac{r}{x}$

$x = \frac{r}{\frac{1}{\sqrt{3}}} = \sqrt{3}r$

$\sin 30 = \frac{r}{y}$

$y = \frac{r}{1/2} = 2r$

Using the formula for the area of a triangle, $Area = \frac{1}{2}ab\sin C$ in conjunction with the formula for area of a circle, $Area = \pi r^2$, we know that:

$Area\ of\ the\ small\ circle = \pi r^2$

$Area\ of\ the\ big\ circle = \pi(2r)^2 = 4\pi r^2$

$Area\ of\ the\ small\ triangle = \frac{1}{2}\left(2\sqrt{3r}\right)\left(2\sqrt{3r}\right)\sin 60 = 12\sqrt{3}\,r^2$

Therefore, the shaded area is: $Shaded\ area = \left(12\sqrt{3} - 4\pi + 3\sqrt{3} - \pi\right)r^2$

$= \left(15\sqrt{3} - 5\pi\right)r^2$

$= 5r^2(3\sqrt{3} - \pi)$

Question 37: A

$(3.12)^5 = (3 + 0.12)^5 = ((3(1 + 0.04))^5 = 3^5(1 + 0.04)^5$

$= 3^5(1 + 5(0.04) + 5\left(\frac{4}{2}\right)(0.04)^2 + \frac{5(4)(3)(0.04)^3}{6} + \ldots$

$= 3^5\left(1 + 0.20 + 0.016 + 0.00064\right)$

$3^5 \times 0.00064 = 0.16$

Therefore, I must obtain 4 terms in the expansion.

END OF SECTION

MOCK PAPER B

Section 2

Introduction:

- Raise the point about how art and culture are hotly debated in the national sphere, and about the possible reasons as to why they should or should not receive funding – Or, in the case of austerity, lesser budget reductions amongst a host of various possible alternatives. Consider your case in light of the possible economic implications of reductions in funding for arts and culture, and other points of view as well.

- The key question: To what extent is art and culture something that deserves public funding, and by whom and for what purpose? What are the potential benefits of funding the arts, and what might the potential tradeoffs be?

Paragraph 1:

- Highlight the economic argument for funding the arts, with a specific example that is of relevance.

<u>- Passage Example:</u>

_A possible reason to support art and culture is that these are economically beneficial industries to a certain extent. The economic impact of arts funding may be direct, insofar as artists who otherwise would not have been able to perform the tasks that allowed them to generate valuable works, or it may be indirect, insofar as these works may have in turn stimulated ideas within others, causing them to create things of value to the economy. Some economic developments may be contingent on arts and culture, or specifically the individuals who are involved in art and culture who would not otherwise be able to take part in a life of art. Art and culture represents a host of ideas and cultural expressions that might not otherwise exist if not for art and culture funding: To the extent that art is considered valuable to individuals and can facilitate the transfer of ideas within a society, one might consider the provision of art and culture funding to be a public good or an investment.

- Example: Artists in Italy after World War II – Italy supported their designers and artists, and the nation was able to bolster its economy by exporting their products, such as Italian leather, etc, which lends credence to the idea that we should support people with the ideas but not the means. National arts grants, that allow artists to make a living.

Paragraph 2:

- Consider including a counterargument – Although the arts can be economically beneficial, it is hard to assess the impact of funding the arts relative to other industries, and this should be factored into the decision of whether to fund or not to fund.

<u>- Passage Example:</u>

Though we see that the arts account for __% of GDP in the current scenario, the impact of retracting arts funding can be hard to measure, and it is questionable who exactly it will impact, in what scenarios. In a situation in which a government must consider questions of what it will cut from the budget, it must think about what is relatively more important to preserve. Other examples: Example of say, a scenario in which a government must debate between cutting pensions vs. cutting art, followed by justifications concerning why it is that one scenario might have more clear cut benefits than the other.

Paragraph 3:

- To question the approach of making the decision of whether to fund the arts or not on the basis of economic considerations, and what the implications of several possible approaches may be.

- Passage Example:

- On the one hand, making the decision to fund or not fund the arts on the basis of economic considerations may potentially be problematic altogether. If a decision to fund art is made on the basis of economic considerations alone, then a government risks funding only very specific forms of art, not others, which in turn may be justification for governments not to focus on the absolute amount of funding that they provide to art and culture in general, but to zero in on the distribution of those funds in particular. On the other hand, perhaps an economic approach to decision-making on behalf of funding the arts is inappropriate altogether, as arts and culture might be, as Letts might argue, more of a moral consideration than an economic one, as the trade-off might in fact be disproportionate in value relative to what economic impact might otherwise suggest.

Conclusion

- Summarize the points:

➢ Art can be treated as something economically valuable to a country, whether directly or indirectly. These might be considered as justifications to fund or not to fund.

➢ On the other hand, assessing the economic impact of that funding is difficult, and Government should assess the potential tradeoffs that it is making when attempting to assess how and whether to fund.

➢ It may be the case that economic considerations may not be the best way to make a decision on whether to fund or not to fund art, as there may be moral considerations that are involved in that process above and beyond our immediate economic ones.

END OF PAPER

Final Advice

Arrive well rested, well fed and well hydrated

The ECAA is an intensive test, so make sure you're ready for it. Ensure you get a good night's sleep before the exam (there is little point cramming) and don't miss breakfast. If you're taking water into the exam then make sure you've been to the toilet before so you don't have to leave during the exam. Make sure you're well rested and fed in order to be at your best!

Move on

If you're struggling, move on. Every question has equal weighting and there is no negative marking. In the time it takes to answer on hard question, you could gain three times the marks by answering the easier ones. Be smart to score points- especially in the maths section where some questions are far easier than others.

Make Notes on your Essay

You may get asked questions on your essay at the interview. Given that there is sometimes more than four weeks from the ECAA to the interview, it is really important to make short notes on the essay title and your main arguments after the essay. You'll thank yourself after the interview if you do this.

Afterword

Remember that the route to a high score is your approach and practice. Don't fall into the trap that "*you can't prepare for the ECAA*"– this could not be further from the truth. With knowledge of the test, some useful time-saving techniques and plenty of practice you can dramatically boost your score.

Work hard, never give up and do yourself justice.

Good luck!

Acknowledgements

I would like to express my sincerest thanks to the many people who helped make this book possible, especially the Oxbridge Tutors who shared their expertise in compiling the huge number of questions and answers.

Rohan

About UniAdmissions

UniAdmissions is an educational consultancy that specialises in supporting **applications to Medical School and to Oxbridge**.

Every year, we work with hundreds of applicants and schools across the UK. From free resources to our *Ultimate Guide Books* and from intensive courses to bespoke individual tuition – with a team of **300 Expert Tutors** and a proven track record, it's easy to see why UniAdmissions is the **UK's number one admissions company**.

To find out more about our support like intensive **ECAA courses** and **ECAA tuition** check out www.uniadmissions.co.uk/ecaa

Your Free Book

Thanks for purchasing this Ultimate Guide Book. Readers like you have the power to make or break a book – hopefully you found this one useful and informative. If you have time, *UniAdmissions* would love to hear about your experiences with this book.

As thanks for your time we'll send you another ebook from our Ultimate Guide series absolutely FREE!

How to Redeem Your Free Ebook in 3 Easy Steps

1) Scan the QR code, find the book you have either on your Amazon purchase history or your email receipt to help find the book on Amazon.

2) On the product page at the Customer Reviews area, click on 'Write a customer review'

Write your review and post it! Copy the review page or take a screen shot of the review you have left.

3) Head over to www.uniadmissions.co.uk/free-book and select your chosen free ebook! You can choose from:

- ✓ The Ultimate ECAA Guide – 300 Practice Questions
- ✓ ECAA Mock Papers
- ✓ ECAA Past Paper Solutions
- ✓ The Ultimate Oxbridge Interview Guide
- ✓ The Ultimate UCAS Personal Statement Guide

Your ebook will then be emailed to you – it's as simple as that!

Alternatively, you can buy all the above titles at **www.uniadmissions.co.uk/our-books**

ECAA Online Course

If you're looking to improve your ECAA score in a short space of time, our **ECAA Online Course** is perfect for you. The ECAA Online Course offers all the content of a traditional course in a single easy-to-use online package- available instantly after checkout. The online videos are just like the classroom course, ready to watch and re-watch at home or on the go and all with our expert Oxbridge tuition and advice.

You'll get full access to all of our ECAA resources including:

- ✓ Copy of our acclaimed book "The Ultimate ECAA Guide"
- ✓ Full access to extensive ECAA online resources including:
- ✓ 2 complete mock papers
- ✓ 400 practice questions
- ✓ Fully worked solutions for all ECAA past papers
- ✓ 5 hours online on-demand lecture series
- ✓ Past Paper Worked Solutions
- ✓ Ongoing Tutor Support until Test date – never be alone again.

The course is normally £99 but you can get £20 off by using the code "*UAONLINE20*" at checkout.

https://www.uniadmissions.co.uk/product/ecaa-online-course/

£10 VOUCHER:

UAONLINE20

Printed in Great Britain
by Amazon